High-Technology New Firms:
Variable Barriers to Growth

High-Technology New Firms: Variable Barriers to Growth

RAY OAKEY

Paul Chapman
Publishing Ltd

Copyright © 1995, R.P. Oakey

All rights reserved

Paul Chapman Publishing Ltd
144 Liverpool Road
London
N1 1LA

Apart from any fair dealing for the purposes of research or private study, or criticism or review, as permitted under the Copyright, Designs and Patents Act, 1988, this publication may be reproduced, stored or transmitted, in any form or by any means, only with the prior permission in writing of the publishers, or in the case of reprographic reproduction in accordance with the terms of licences issued by the Copyright Licensing Agency. Inquiries concerning reproduction outside those terms should be sent to the publishers at the abovementioned address.

British Library Cataloguing in Publication Data
Oakey, R. P.
  High-technology New Firms: Variable Barriers to Growth
  I. Title
  658
  ISBN 1-85396-239-2

Typeset by Hewer Text Composition Services, Edinburgh
Printed and bound by The Cromwell Press Ltd, Broughton Gifford, Melksham, Wiltshire

A B C D E F G H   9 8 7 6 5

CONTENTS

1	**Introduction**	1
	Contextual issues	1
	Background	1
	Some fundamental misconceptions	3
	Objectives	4
	The survey	6
2	**Theoretical and Practical Contexts**	9
	Determinism and industrial strategy	9
	Determinism in science	9
	Technological determinism	11
	Technological determinism and new high-technology small firm formation	12
	Contextual formation attributes of survey firms	14
	Founding stimuli	15
	Management structure of the new firm	16
	The sectoral origin of the main founder	17
	Links with the previous employer	21
	Premises	22
	Summary conclusions	26
3	**The Interacting Impacts of the Founding Product Technology and Funding on Firm Formation and Growth**	29
	Product innovation behaviour, new firm formation and growth: some introductory conceptualisations	29
	Simplified models of new firm formation and growth	30

	Problems associated with "near market" exploitation	33
	The derivation, mix and financial contribution of founding technologies	35
	The origin of the founding product technology of the new firm	35
	The market launch of the *founding technology*, and its contribution to the financial viability of the firm	36
	The unit profit margins achieved by the initiating product or service technology	38
	Changes in post-formation product mix	41
	Areas of future production growth	42
	The general funding of formation and growth	43
	Funding at the time of formation	46
	The contribution of external funding *three years after formation*	47
	The contribution of profits *three years after formation*	48
	The involvement of external investors through equity shares	50
	Use of a formal business plan at the time of formation	52
	The current financial position of the firm	55
	Evidence on the overall year end profitability of the firm in 1991	57
	The extent of additional external funding for specific projects in established firms	58
	The funding of future growth	60
	Summary conclusions	61
4	**The Role of R&D in High-Technology Small Firm Formation and Growth**	**65**
	Introduction	65
	The extent of R&D commitment in survey firms	68
	Research and Development linkages	72
	R&D sub-contracting for external agencies	72

	Research and Development contracts awarded to external agencies	74
	Important sources of external technical information	77
	Other forms of technology acquisition	77
	Employment problems and internal innovation	80
	Summary conclusions	81
5	**Purchasing, Sales and Marketing**	**81**
	Introduction	81
	Purchasing patterns	84
	Input materials	84
	Sub-contracting	87
	Customer patterns	89
	Incidence of a single major customer industry	90
	Incidence of a single major customer firm	91
	Exports	93
	Sales and marketing	93
	Main product promotion methods	94
	Commitment of resources to sales and marketing	95
	Marketing agreements	98
	The competitive environment	99
	Summary conclusions	99
6	**Acquisitions**	**103**
	Introduction	103
	The incidence of acquisition in survey firms	105
	Acquisition criteria	107
	The motivation for acquisition	107
	Previous association with acquirer	108
	The value of the acquisition to acquired firms	108

	The strategic relevance of acquisition to survey firms that remain independent	109
	Attempts to acquire other firms	109
	Unsuccessful acquisition attempts towards independent survey firms	110
	The envisaged strengths and weaknesses of acquisition to independent firms	111
	Summary conclusions	112
7	**Conclusions**	**115**
	A summary of results	115
	Implications for theory	117
	Implications for policy	122
	Variable sectoral activity: implications for policy	122
	The broader picture	125
	Bibliography	**127**
	Index	**131**

Acknowledgements

I would like to thank several individuals who have assisted with the fieldwork on which this book is based, and with the final preparation of the text. In chronological order, acknowledgement is due to Sarah Cooper and Janet Biggar for their assistance in deriving a sample of firms and collecting the data on which the empirical findings are based. I would also like to thank Tim Watts, Andrew James and Francis Chittenden for comments on earlier drafts of the text. Further acknowledgement is due to Jane Geddes and Kate Todd for their typing assistance in the final stages of preparing the manuscript. However, as usual, any errors or omissions contained in this work remain my responsibility. A final general acknowledgement is due to the SERC/ESRC Joint Committee for the funding of the research on which this book is based.

CHAPTER 1

Introduction

Contextual issues

Background

It is regrettable that much of the enthusiasm shown by politicians and media "experts" for technical subjects is, rather perversely, often triggered by an early, and consequently, over-simplistic interpretation of the phenomena concerned. There is a sense in which unsubstantiated and misleading scraps of information can be accepted as proof of an assertion, *simply because* no detailed evidence exists with which they might be refuted. Usually, when subsequently explored by means of rigorous scientific method, the real picture is far more blurred and complicated than the unrealistic (yet seductive) state of affairs originally speculated. The recent controversy over "cold fusion" in physics is an illustration of this phenomenon, where early claims are reported as facts, while later complicating detailed evidence is ignored as the media moves on to the next "breakthrough". An enthusiasm for New Technology-Based Firms (NTBFs) throughout the 1980s in Europe, as vehicles for rapid industrial employment growth, is a further example of how a simplistic proposed solution to the problem of traditional industrial decline can be accepted, virtually without question, as a viable panacea. Such irrational behaviour owes much to two powerful stimuli. First, in keeping with the above assertions, evidence available to support the NTBF growth case was either not directly relevant, or totally absent. For example, most of the evidence available on the growth of NTBFs in the early 1980s, when their popularity was at its height, was of effects (i.e. employment growth), rather than causes (Morse 1976), being based on data from the United States. Indeed although, in an American context, the *end result* was clear in employment growth terms, the causality of how it had been achieved and, perhaps more importantly (once identified), *whether* it could be replicated in a European context, were issues that were not considered. Second, such unsubstantiated enthusiasm was partly triggered by the *urgent* political need to solve the intractable problem of rising industrial unemployment.

In the radically changed political and economic climate of the early 1980s where, in both the United Kingdom and the United States, extreme "free market" principles had been recently adopted, the twin entrepreneurially-oriented panaceas of new small firms and emerging high-technology industrial sectors were perfectly focused in the form of NTBFs. Rapid employment growth in the small firms sector in general (Birch 1979) and high-technology small firms in particular (Rothwell and Zegveld 1982) appeared to suggest that NTBFs would ideally exploit these new non-interventionist "free markets", because of their inherent high profitability that would negate the need for state subsidy which had been necessary to support the "problem sectors" of the 1970s. Such older industries, typified by iron and steel, textiles, and more recently, consumer electronics and motor vehicles production, might be largely abandoned to exploitation by emerging third world countries as developed western economies retreated into the R&D intensive "start of cycle industries" (Rothwell and Zegveld 1985), where competition was less severe and profit margin wider.

Certainly as far as the United Kingdom is concerned, as we approach the year 2000, these enthusiasms have proved to be ill-conceived. While the United Kingdom has continued to lose the traditional industries that led the post-Second World War recovery, there has been little compensating growth from the new high-technology industries in either large or small enterprises. While the aggregate growth of NTBFs has been unspectacular (Keeble and Kelly 1988; Oakey 1991), a series of "winner", fast-growing NTBFs have appeared, only to disappear within a few years, beginning with Sinclair Research, followed by Acorn Computers and ending with the sale of INMOS to the French Thomson CSF group. In terms of large United Kingdom high-technology firms, the 1980s ended with the sale of ICL, the only large United Kingdom mainframe computer manufacturer, to Fujitsu of Japan. Other large high-technology manufacturing firms (notably GEC) are heavily dependent on the defence industry for much of their high-technology market, in a period when "peace dividend" cut backs imply an increasingly competitive and reducing market for high-technology defence hardware.

A rather cavalier attitude to the understanding of how high-technology industry works lies at the heart of any explanation of this sad state of affairs in the United Kingdom. Initially, the assumption that existing NTBF growth, evident in the United States, could be simply replicated in Europe in general, and the United Kingdom in particular, was simplistic in the extreme. The direct and indirect importance of massive defence industry expenditure, more liberal taxation regimes, and a more aggressive entrepreneurial approach to business (especially in higher education), are only three of many reasons why the NTBF growth of the United States might not be replicable in Europe, where conditions are sharply different in many countries. Indeed, the European Science Park movement, in particular, was based on the largely false assumption that proximity to universities would be a major stimulation to NTBF formation and growth. Although the founders of such developments often stressed the potential for technical interaction between Science Park firms and university staff, as a major trigger for NTBF growth, accumulating evidence showed that in Silicon Valley, a region frequently cited as being an example of this phenomenon, such links between NTBFs and Stanford University were weak (Oakey 1985). Ironically, factors more important in explaining United States NTBF growth of the Silicon Valley type, in particular lower taxes (especially on capital gains), and greater academic entrepreneurship, largely remained unaddressed by the founders of European Science Parks.

Another major feature of NTBF formation and growth in Silicon Valley and other high-technology agglomerations in the United States was the rather obvious fact that none of them were particularly new! The belief in the early 1980s among many European pundits that a series of European Silicon Valleys could be developed before the end of the decade flew in the face of all the United States evidence. If Silicon Valley is again taken as an example, by the early 1980s, the agglomeration that then existed had been growing since (at least) the early 1950s; clearly implying that any attempts at replication should be medium to long term in nature, and could not be a short-term "quick fix" to regional unemployment problems. If the above mentioned point that proponents of Science Parks in Europe were stressing erroneous growth stimulation factors, while ignoring the key causes of growth, the prognosis for European Science Parks was not encouraging.

Indeed, after a decade of Science Park promotion in the United Kingdom, there have not been any startling success stories. Although the Cambridge Science Park is well established, there is no real evidence that embryonic replica Silicon Valleys have been created anywhere in the United Kingdom. Growth in most of the Science Parks formed throughout the 1980s has been slow, or has stagnated. In the absence of any strongly beneficial technical links to be gained from a Science Park location, the frequently high Science Park rents, when compared with other local locations, often ensures that high quality premises, that convey the "right image" and a "good address" (e.g. Cambridge) are the only tangible benefits of a Science Park site. In the case of Science Park policy, the provisions that have been made are more representative of the need for short-term actions in response to local political needs, than a considered attempt at addressing the formation needs of NTBFs as a means of improving their potential for survival and enhanced growth.

Some fundamental misconceptions

Probably one of the most serious misconceptions concerning high-technology industry in general, and NTBFs in particular, is the belief that the measurement of R&D *inputs* could act as a surrogate for high-technology growth, or growth potential within national or regional economies. The observation that most high-technology firms spend a considerable amount of their available capital on R&D has, rather illogically, led to an assumption that such industries are desirable vehicles for achieving rapid private sector industrial growth. While in a number of limited, rather over-publicised cases from the United States, this is true (e.g. Apple Computers), in most cases, the use of an operating *cost* as measure of success is a particularly risky practice! In both the United States and the United Kingdom, this tendency has been partly vindicated by the large proportion of the high-technology industries of these countries that have traditionally, in whole or part, been engaged in defence industry manufacturing, where cost-plus contracts and unrealistic final prices for products have meant that, to some extent, high R&D input costs have been synonymous with high profitability. However, in the largely commercial world in which NTBFs operate, high R&D input costs are only a desirable attribute if they produce high value R&D *outputs* in the form of competitive commercial product sales. Unfortunately, it has been noted elsewhere that, although R&D outputs in the form of product sales would be a more accurate measure of high-technology success, the aggregate (often government) statistics used by high-technology industry researchers tend to be R&D input oriented (Oakey et al 1988; see Chapter 4).

All the above assertions, and accumulating academic evidence from a wide range of recent sources (Oakey 1994), suggest that most NTBF formations are not characterised by fast growth, and are consequently not realistic vehicles for ameliorating sharp industrial decline in other industrial sectors. However, this observation should not be judged to indicate that NTBFs are, therefore, of no significance to the future industrial growth of industrialised countries. Certainly in the United Kingdom, government interest in NTBFs has ebbed and flowed in a rather illogical manner over the past fifteen years partly, as discussed above, due to a lack of hard scientific evidence on which judgements could be based. In the early 1980s, when little clear evidence existed on the mechanisms that influenced NTBF growth, United Kingdom government interest in NTBFs was very strong. It is now perversely true that the recent generation of evidence on the problems of NTBF formation and growth by researchers has coincided with a decline of government interest in such problems.

This erratic support for NTBFs is inconsistent and wasteful. Although most NTBFs do not have fast growth potential, they often represent a key source of technical innovations in the sectors of which they form a part. Their capacity for radical innovation that may trigger a new pulse of growth in a different technological direction from within an existing high-technology sector is well known. For example, NTBFs were largely responsible for the formation and early growth of the semiconductor industry in the United States (Freeman 1982). Indeed, experience from the semiconductor industry has shown that, given conducive economic conditions, NTBFs can grow both individually, and in aggregate, to make a substantial contribution to national sectoral employment in the long term. However, any view remaining from the early 1980s that NTBFs were largely independent, and would not require substantial external public support, has transpired to be unrealistic. Due to the frequent need of NTBFs for substantial amounts of "front end" investment to fund essential R&D over protracted periods of time, substantial external support is required in many cases. Although the risks of whole or partial failure always accompany any high-technology venture, the rewards of making a moderate or radical technological breakthrough are also high and approximately proportionate.

Objectives

Seen in a context of high risk, any information that reduces uncertainty concerning NTBF formation and growth is of value. The research on which this book is based was performed because it is argued that existing definitions of NTBFs tended to be general "caricatures" upon which both public and private sector external investors in NTBFs have been basing their decisions. Anecdotal evidence from unrelated earlier work (Oakey 1984; Oakey et al 1990) convinced this author that substantial functional differences existed between a range of firms that would all be generally classified as NTBFs with fast growth potential. Without making allowances for these real differences in terms of, for example, product development "lead times" and points at which "break even" from sales could be achieved, external investors, with the power to assist such new enterprises financially, have become confused. In the absence of clear evidence, they have developed either grossly over- or underoptimistic views of what can be achieved by an NTBF in a given sector. The study on which this book is based seeks systematically to analyse the formation and growth characteristics of new firms in the biotechnology, electronics and software sectors of industry, in order to measure initially *if* substantial differences exist between these sectors, and if they do exist, to record their nature and their extent. Clearly, a detailed understanding of differences between sectoral groupings of NTBFs, and why these occur, aids our theoretical understanding of high-technology small firms. However, this information has additional clear implications for policy development, since the generation of such evidence will strongly influence what external investors can expect to receive, and what they should expect to give when they are contemplating an investment in an NTBF of any given sector.

This intended clarification is attempted through the following chapters that present the main evidence of the book. In Chapter 2, the deterministic nature of many NTBF formations is hypothesised as a conceptual theme for the whole book. This approach examines both the theoretical and practical ramifications of the constrained product or process choice for new firm formation, that confront NTBF entrepreneurs. This fundamental theme of limited choice of technological basis for an NTBF, that is determined by the development of specialist technical expertise by entrepreneurs, is returned to in the conclusions of the book in Chapter 7.

Chapter 3 is concerned with a major investigation of the formation and related funding process in NTBFs. This is the largest chapter in the book, and considers the portfolio of funding methods that are utilised by the firm during the first year after formation, and the evolution of funding in the years following "start up". Many of the funding problems at the time of formation stem from long "lead times" on product development. Thus, the financial problems experienced at the time of formation are appropriately followed by a detailed consideration of R&D in NTBFs in Chapter 4, since it is this form of technically focused investment that causes most of the financial difficulties within NTBFs, and high R&D spending is the factor that distinguishes such firms from most other lower technology forms of production.

In Chapter 5 the often geographically and sectorally complex nature of the survey firms' purchase, supply and marketing relationships are examined. The customers of low-technology firms are often local, since the relative ubiquity of their products ensures that sales to distant locales are unlikely to be successful due to identically competitive products. For example, it is rare for a small scale British printed circuit board maker to sell boards in Germany, a location where other local producers would have many cost advantages. However, this assertion does not apply for many NTBFs' products in this survey, where the unique specification of product designs ensure that international markets can be penetrated, while such sales are rendered economically viable by the high profit margins that are chargeable.

The analysis of this book is concluded in Chapter 6 by an investigation of acquisitions that have occurred in a minority of firms *since* formation. This chapter should be viewed in the context of the results of Chapter 3 on formation and funding, since acquisition is a radical funding option for independent firms in need of investment capital. While, in specific cases, a buyer may be sought by the founder of an NTBF when the firm concerned has reached a point at which its sale is advantageous to the founder, most acquisitions are an option of last resort for many NTBF founders. Indeed, a major motivation for firm formation (noted in Chapter 2) is the desire to obtain independence, often from a previous large firm employer. The use of acquisition as an unwelcome, but often sole available, means of obtaining financial support re-evokes that problem of funding NTBFs, which is an issue that is central to many of the empirical sections of this book. Thus, it should be of no surprise that funding problems, together with other major findings of the empirical chapters, are discussed in the conclusions of Chapter 7, with particular regard to their relevance to theory and policy towards NTBFs.

The survey

In advance of the following chapters that utilise evidence from the field survey in support of arguments on given topics, an explanation of the survey design is warranted in order to set such data in context. As noted above, there existed at the time of study, considerable fragmentary evidence from earlier studies that the proposed homogeneity, attributed to high-technology industry, by those writers attempting general definitions (Breheny and McQuaid 1987; Thompson 1988), was unlikely to exist in practice. However, there was no consistent data set that emanated from a controlled analysis of formation and growth variations between sectors, generally acknowledged to qualify as high-technology forms of production. The survey on which this book is based sought to fill this gap in our knowledge by conducting such a systematic study of formation and growth in four high-technology sectors: materials technology, biotechnology, electronics and software. It transpired, after considerable work on directory sources, that although viable universes of new firms could be derived for biotechnology, electronics and software, only eight valid materials technology firms could be identified. Due to the evolution of this sector, most of the firms identified as materials technology based were invalid, either due to their subsidiary status (i.e. established as a subsidiary of a large existing firm), or their age (older than the maximum age adopted for the study − see below). However, considerable initial investigative work took place before these exclusions could be made.

Indeed, a major problem in conducting research into new small firms in given industrial sectors is the dearth of the required *accurate* information from general directory sources on formation date (i.e. in this case, the decade between 1978 and 1988), productive activity (i.e. goods and/or services in biotechnology, electronics or software production), and ownership (i.e. totally independent when founded). In order to reduce this problem, the research was designed in three stages, beginning with a comprehensive search of all relevant directories to include in the survey any British firm that appeared to meet the above age, sectoral and ownership characteristics. This first stage of universe generation produced 2,237 potential study firms.

The second stage of the process involved sending all these firms a brief postal questionnaire to gain accurate information on the above key criteria. The 987 respondents represented a respectable 44.12% response rate to this postal survey. The final stage in the methodology was the random approach of 50 firms from each of the three sectors chosen for this study. This objective was broadly achieved by interviews conducted in the sectors of biotechnology (44 firms), electronics (46 firms) and software (47 firms). The overall response rate to the interview survey was a very high 91.33%. These data form the basis for the following discussion and analysis. Although in the following analysis, member firms of the three study sectors will be aggregated and represented as a group, Figure 1.1 initially gives a more detailed breakdown of the wide variety of activities that exist in all three study sectors when disaggregated to the level of the enterprise.

Figure 1.1 Breakdown of activities by sectors

BIOTECHNOLOGY

			N	%
Human health	-	diagnostic kits marketed to hospitals	12	27
Services	-		11	25
Biologicals	-	reagents for use by others. Intermediate products	5	11
Equipment/Inst	-	fermentation/lab equipment used in biotech environment	5	11
Chemicals	-	reagents	3	7
Foods	-	gelatin, vinegar	3	7
Animal Agriculture	-	cow breeding & diagnostic kits	3	7
Environment	-	pollution control for manufacturing techniques for control of effluent from fermentation process	2	5
	Total		44	100

ELECTRONICS

			N	%
Printed circuit boards (PCB)				
TV/video	-	specialised closed circuit TV for security purposes. Range of small units. Also police double recording cassette systems.	5	11
Service	-	design systems to other specifications	4	9
Power	-	small generators, BT mobile units	4	9
Comms	-		3	6.5
Instrumentation	-	control	3	6.5
Other components	-	capacitors (some large for power stations)	8	17
Other equipment	-	Finished products for power & electronics	13	28
	Total		46	100

SOFTWARE

			N	%
Specialist software	-	packages e.g. for motor traders & plumbers i.e. for particular end-users	13	32
Financial/Accounting	-		8	19.5
Bespoke	-	customer design, client specific as existing packages insufficient/unsuitable	7	17
Utilities	-	Norton utilities. Virus checker - fit operation	3	7
Management systems	-		3	7
Other business software	-		3	7
CAD	-		1	2.5
Other	-	non-specialist wordprocessing, also training consultants for transition to software group are not specialist business nor utilities, more general business support.	3	7
	Total		41	100
	Grand total		131	

CHAPTER 2

Theoretical and Practical Contexts

Determinism and industrial strategy

Determinism in science

The issue of determinism has a long and well established history within philosophy of science literature (Harvey 1973). The fundamental question whether man is in control of his actions, or whether his behaviour is determined by a plethora of external forces, lies at the heart of principles concerning the law, religion and the social norms that often manifest themselves in terms of political ideology. For example, while those who believe in "free will" may argue that a criminal is totally responsible for his or her actions (and must consequently suffer retribution), others, taking a more deterministic view, might argue that such individuals are merely victims of a previous deprived environment that has largely determined such behaviour, implying the need for rehabilitation rather than crude punishment.

The heritage of determinism largely stems from the approach of many mid-nineteenth-century physical and social scientists who believed that nature, and the human condition within this overall regime, was evolving under a predetermined plan, and that the role of science was to discover the basic mechanisms of this "grand design". It was argued that, once the mechanisms and directions of such trajectories were discovered, the future course of various phenomena could be broadly predicted. The work of Sir Charles Lyell within geology and the dialectic view of Karl Marx can be seen in these terms. However, the historical determinism of Marx was criticised by Popper (1965), not merely for the constraining influence such a predetermined approach exerted on later neo-Marxist sociologists and economists, but more importantly, for the inability of neo-Marxists to reject the precise predictions of Marx, once it was clear that his prophecies were not realised as the twentieth century unfolded. Moreover, other twentieth-century philosophers have questioned a fundamental principle of determinism by arguing that the basic assumption of a "grand plan" and the implicit belief in progress, is invalid, and that a less value-judgement laden concept of "change" should replace "progress" when discussing the way in which science evolves (Kuhn 1970).

A major danger inherent in all arguments that are based on determinism is the fatalism that stems from an acceptance that there are laws external to human control that determine events. Once individuals accept that they are not able to shape events, it becomes easy for them to divest themselves of responsibility for their behaviour. This phenomenon is exemplified by the rise of environmental determinism within geography at the turn of the century. Central to this approach was the assertion, based on superficial circumstantial evidence, that peoples' competencies were determined by their environment. Following the theory, Latin peoples were "hot blooded" and lethargic due to the hot climate which made them irritable and lazy. Indeed one, now infamous, American geographer claimed that Negroes would not be able to migrate to the northern parts of the United States due to their inability to survive cold winters (Semple 1911)!

On a more serious note, however, much of the philosophical basis for the racist views of pre-Second World War Nazi Germany was derived from a deterministic interpretation of the work of geographers on environmental determinist, and a mis-interpretation of the work of Charles Darwin on the origin of species in which the *random* selective basis for evolution was ignored, while evolution, and in particular a "survival of the fittest" approach, was used as a justification for a claim for racial supremacy. Again, a crucial danger of this determinist approach is the accompanying implication that no moral responsibility attaches to the individual, since we must surrender to the dictates of a process of development that is determined.

At an individual practical level, however, our everyday attitude to determinism is largely influenced by our knowledge-base. While in principle, given access to an aircraft, we might have the option of piloting a solo flight (implying the exercising of free will), our propensity to attempt a solo flight would largely depend on our previous flying experience. Put simply, in most "real world" environments, although the option of making a decision not influenced by experience is possible, most individuals will choose the option, based on their experience, that will deliver the best perceived chance of solving the problem with which they are confronted. Indeed, most theories in the social sciences are based on probability rather than determinism (Harvey 1973), since they are an attempt to *predict* in probabilistic terms how human beings will behave under certain conditions.

The aggregated actions of individuals, when expressed as social behaviour, may be expressed as a social theory of human behaviour. In this context, as noted above, the deterministic approach of Marx was criticised by Popper not on the grounds that it was deterministic (since most scientific theories are of this type), but because neo-Marxists refused to reform the theory in the light of evidence that its predictions were incorrect (Popper 1965). Neo-Marxists generally made the mistake of either refusing to accept the decision of the millions of individuals who took the non-Marxist option, or claiming that the Marxist solution *would* occur, given time. The tendency to refute conclusive scientific evidence that a theory is wrong, and maintain a theory, is another worrying side effect of a strongly deterministic approach in which a theory takes on almost a *religious* significance in circumstances where belief in a principle no longer needs to be supported by scientific evidence (e.g. racial supremacy), or more to the point, cannot be refuted by clear scientific evidence that shows it to be false.

Technological determinism

Although the above discussion on determinism might appear to have little relevance in an industrial context, the continuing strength and relevance of this issue to all aspects of social science research is reflected in the vigorous debate that continues in the field of industrial management concerning technological determinism. Indeed, the fundamental principle of whether management can be freely determined, or is the (at least a partial) prisoner of technological imperatives, is very similar to the other general examples of free will versus determined action discussed above.

Perhaps the significance of technological determinism in the post-war period has stemmed from the realisation that technological change has increasingly become a major explanatory factor in determining the success of individual industrial firms (Mansfield 1968; Freeman 1982), regional economies (Thomas 1975; Thwaites 1978; Oakey 1984) and levels of national performance (Solow 1957; Denison 1967). Through the application of new technologies in terms of processes, products and management devices, industrial firms have expanded the production possible from given limited units of capital and labour to levels of output unimagined by Victorian economists (e.g. Marx). In many sectors, a failure to maintain high levels of investment in "state of the art" process and related product technologies has led to the demise of whole industries. Unfortunately, Britain in the post-war period has experienced the whole or partial demise of a number of industries that illustrate this phenomenon (e.g. textiles; shipbuilding; motor cycles; machine tools; televisions) (Rothwell and Zegveld 1981).

It is certainly the case that large-scale capital intensive industries are particularly sensitive to the rapid pace of technological change to the extent that a failure to adopt new process technologies, or even a delay in their adoption, can cause the collapse of the complete sector within a national or regional economy. For this reason alone, it is not surprising that seminal work on technological determinism by Woodward (1965) stressed the importance of production technologies in providing a key deterministic influence on the successful organisation of industrial enterprises. While the arguments of Woodward were couched much more in terms of emphasis than unequivocal assertion, her approach tended to stress that technology determined the appropriate method of production, rather than strategic management determining the choice of production method, particularly with regard to appropriate process technologies at different *scales* of production.

Other subsequent researchers have criticised the Woodward approach and have chosen to stress the importance of management freedom in *selecting* the technology to be adopted by the firm. Such an approach tends to depict technology as a tool of management to be used as part of the overall strategy of the firm, but significantly not as an essential prerequisite for survival in a deterministic manner (Child 1972). This alternative approach tends to emphasise the role of "strategic choice" in determining the success of the firm, and is based on a wider interpretation of the firm environment to include, not only the production methods of the firm and its workers, but also the complete entity of the enterprise from which the overall performance of the firm is derived (e.g. R&D, marketing, administration etc.). There is more emphasis in the work of Child on the decision process prior to the adoption of technology, rather than the Woodward approach of stressing strategic behaviour after adoption.

Here again, the classical debate of determinism is apparent in that Woodward emphasises the deterministic constraints of technology, while Child stresses the "free will" argument by emphasising the role of human choice in terms of control of the technology selection process within industrial organisation. However, it is often the case that extremes of debate within an academic discipline are both partially correct, and that there is a "mid ground" between thesis and antithesis that is a workable compromise. As mentioned above, there are areas of production within industry where technological determinism is very strong. Apart from specialist "niche" production (e.g. Morgan Cars), batch production in the motor vehicle industry would generally not be an advisable strategy for competition in the high volume car market, where capital and subsequent process technology investment is high.

Indeed, Woodward admits that technological determinism is most applicable to such mass production forms of manufacture. She suggests that, within jobbing and batch production, there is far greater scope for management choice since there is no well established highly efficient production paradigm, and both process and product technologies are far more volatile. However, even in areas of mass production, there continues to be scope for management choice because technology is rarely static and importantly, can only be deterministically constraining if there is no scope for change. For example, the recent introduction of "Just in Time" (JIT) methods of production have revolutionised the already capital intensive high-technology manufacturing processes of the motor vehicle industry. Such an example offers proof that technological determinism may not be all constraining *if* new managerial choices are able to improve production efficiency by changing the nature of the technology, or the manner in which various aspects of that technology are organised. Conversely, if strategic choice is merely concerned with the exercising of an independent approach that makes no economic sense, and is inferior to best technological practice, it is likely that such an indulgent approach will not meet with success, and will probably cause the demise of the firm concerned.

Technological determinism and new high-technology small firm formation

Although possibly not immediately obvious, the above brief introductory discussion on determinism in general, and industrial determinism in particular, has strong relevance to new high-technology firm formation and growth. While in many ways, the decision to begin a new small firm is often based on a desire to escape the bureaucracy of large employer organisations, the freedom of action available to new firm founders is often less than initially might be expected. At this point, the above discussion of the importance of experience in determining the advisability of choosing a radical course of behaviour is again relevant. This observation is particularly germane to the formation of new high-technology small firms, since it is virtually impossible for a new firm founder to establish such an enterprise without substantial previous technical experience. While it is theoretically feasible for a businessman with substantial surplus capital, and no technical experience, to begin a new high-technology business, most new high-technology small firm founders are technical entrepreneurs (Rothwell and Zegveld 1982; Oakey 1984; Roberts 1991). This phenomenon is explained by the reality that it is difficult for any entrepreneur to be a "product champion" for a new technology on which a firm's formation is based, if he or she does not have an *intimate* working knowledge of the technical potential of the proposed product. Technical entrepreneurs are prevalent as firm founders because they possess both a full understanding of, and commitment to, the new technology concerned, together with the subsequent personal drive necessary to make it a reality in terms of a new firm formation.

However, such ability is not won cheaply in terms of personal experience, and it is a basic assertion of this book that, in a perverse manner, the otherwise advantageous experience of new high-technology firm founders is a deterministic constraint on the type of business that might be formed. Experience in studying the formation and growth of high-technology small firms over a number of years has shown (Oakey 1981; 1984; Oakey et al 1988) that there are only two major sources of new high-technology small firm entrepreneurs. First, the substantial technical expertise necessary to form such a firm can be provided by a higher education institution, usually in the local area of the new firm. It is the *local* "spin off" of such technical entrepreneurs that largely explains the existence of high-technology industrial agglomerations around Stanford University (Silicon Valley), Massachusetts Institute of Technology (Route 128 [Boston]) and the Science Park near Cambridge University in England (Cooper 1970; Oakey 1985). Second, it is now well established that other high-technology small firms are formed as a result of "spin offs" from existing large (or at least well established) firms in such agglomerations. Indeed, in an area such as Silicon Valley, *both* academic *and* industrial "spin offs" have been contributors to the growth of the subsequent agglomeration (Freeman 1982; Rothwell and Zegveld 1982).

It is a major assertion of this book that the technological basis for most new firm formations will have been established in either an academic or industrial establishment in the years prior to the formation of the new firm (Roberts 1991). It is further argued that a major reason for the tendency for many new entrepreneurs to be aged between 30 and 40 at the point at which the new firm is established, is that it is only after ten to fifteen years that a potential entrepreneur has developed sufficient technical expertise on which a new enterprise might be based. However, the essence of the deterministic function grows with increasing specialisation such that, with only rare exceptions, a gifted biologist will form a new firm in the biotechnology industry, while an electronics engineer is determined to enter the electronic industry by virtue of their specialist skills. While it is *theoretically* possible for an electronics expert to begin a new business in the software industry, his best chance of developing a product with a competitive edge, and in attracting external financial support, is on the basis of his accumulated expertise. Put simply, increasing expertise has a growing deterministic influence by reducing the choice of individuals with regard to the sectors in which they can realistically begin businesses.

Thus far, it might be argued that the above comments, although (possibly) interesting from an academic viewpoint, have little policy or practical relevance. If it was the case that barriers to entry in differing high-technology sectors were of equal severity, then the sector in which new high-technology firms were formed would have little significance, since a general policy would suffice all. However, a further major hypothesis of this book is that barriers to entry and subsequent growth *are not equal*. If this assertion should prove correct, it therefore follows that, not only are entrepreneurs largely determined in their choice of the sector in which they launch a new firm, but also the problems they face at formation, and in the subsequent early years after establishment, may vary sharply *between* sectors.

If it is proved correct that high-technology new firm formation will vary in terms of difficulties they experience as a result of sectoral formation "choice" within high-technology industries, *and* that such difficulties are largely derived from factors associated with the nature of the industry that are outside the founder's control, there are important implications for the founders, external funders and government supporters of new high-technology small firm ventures. It is a fundamental thesis of this book that the growing disenchantment of British venture capital agencies with new high-technology small firm ventures over the decade of the 1980s has largely stemmed from ignorance of high-technology industries in general and the variable formation and early growth requirements of differing forms of NTBF in particular. It is hoped that this book will conclusively prove that "pay back" periods for different high-technology small firms in different high-technology industrial sectors will vary widely and should be taken into account when funding individual high-technology ventures. While there remains some scope for gifted entrepreneurs to show a degree of "free will" in effectively managing new high-technology ventures, basic features of the industrial sectors entered by such founders including, in particular, variable "lead times" on initial products, are determining technical constraints that *must* be built into any evaluation of individual business plans. A failure to realise this reality may mean that a new firm venture that had good medium to long-term potential is not supported due to a superficial evaluation, based on a short-term assessment of growth potential that does not pay sufficient attention to the technologically-based development needs of the firm. This would be a loss to the firm concerned, the potential external funder and the wider industrial economy to which it might have made a valuable contribution.

Contextual formation attributes of survey firms

The preceding parts of this chapter have made a number of assertions on factors that determine the scope for firm formation choice among new high-technology small firm entrepreneurs. As an initial stage in a continuing process developed in subsequent chapters, this sub-section will utilise evidence from the survey of new firm formations in the biotechnology, electronics and software sectors to test a number of the arguments made in relation to the constraints on new firm formation and growth. In particular, if the deterministic arguments advanced above are accepted, there should be strong measurable respective links between the technical and geographical origin of the previous employer and the technical basis and founding location of the new "spin off" firm.

Founding stimuli

An initial key question to survey respondents on the founding process concerns the main reason for the establishment of the business. While it should be acknowledged that any decision to begin a new firm depends on many partial stimuli, there is some value in establishing the *major* trigger for the formation.

Interestingly, the single largest reason for new firm formation was the existence of a new product idea that founders believed their new firm could exploit. Almost half (i.e. 55 [41%]) of responding firms to a question on the major reason for foundation stated a new product idea was the stimulus (Table 2.1). This is early implicit evidence that new firm founders are strongly determined by their previous sectoral activity in that, for both industrial and higher education origins, founders largely develop new product ideas based on work they were performing in their previous employment. Thus, a lecturer and/or a researcher in a computer science department at a university is more likely to begin a new firm in the software industry than in electronics. This result supports other previous evidence which suggest that a new product idea, and the desire to gain personal intellectual property ownership, is a major stimulus of many new firm formations (Oakey et al 1988; Roberts 1991).

TABLE 2.1 *Reason for formation by region*

	Rest of Britain		South East and East Anglia		Total	
	N	%	N	%	N	%
Freedom	19	26.8	27	42.9	46	34.3
Redundancy	3	4.2	8	12.7	11	8.2
Product exploitation	35	49.3	20	31.7	55	41.0
Other	14	19.7	8	12.7	22	16.4
Total	71	100	63	100	134	100

Chi Squared = 8.95 P = 0.03

Another major reason for the formation of survey firms was the desire among founders for the greater freedom that owning their own business would bring. This was the main motive for foundation in 46 (34%) of responding cases. The only "push" factor noted as a rationale for the founding of a new business was redundancy, which was cited by a minority 11 (8%) of respondents. Interesting statistically significant regional difference during Chi Squared testing at the p = 0.03% level was produced in Table 2.1, by comparing firms in the South East and East Anglia with the rest of Britain. While freedom was a much stronger motive in the South East in 43% of respondent firms, the equivalent figure for the rest of Britain was 27%. Conversely, the more practical rationale of the existence of a new product idea was more popular in the rest of Britain in 49% of cases, compared with 32% of Southern respondents. This more "hard nosed" approach of founders in the rest of Britain might reflect a comparatively more difficult environment for new firm formation and growth in which new enterprises are only contemplated if there is a "concrete" product idea on which to base a new venture. However, the greater incidence of redundancy driven formations in the South East and East Anglian sub-group of firms may reflect the late 1980s recession, that was "biting hard" in the South East in the period up to (and including) 1991, and had not, by then, fully reached the rest of Britain.

Management structure of the new firm

A very strong result that might, to some extent, confound venture capitalists who are often keen on supporting a *team* of high-technology new firm founders is the clear evidence from Table 2.2 that rapid firm growth (represented here by turnover growth) is strongly related to "single founder" businesses, of sufficient difference to produce a Chi Squared test significant at the $p = 0.001\%$ level. At the extreme, it is clear that, while only 10% of single founder businesses fell in the less than 25% growth category, the equivalent figure for businesses with four or more founders was 33%. However, while there is clear evidence in Table 2.2 of the impressive performance of single founder businesses, it should be remembered that all the firms in this study were less than ten years old at the time of survey. In the medium to long term the literature suggests that the rather autocratic forms of management common in single founder businesses may become less appropriate as the delegation of key functions becomes necessary (Roberts 1991). An unwillingness to delegate can cause constraints to growth that might change the pattern of results achieved in Table 2.2, given time. It is also worth noting with regard to venture capitalist involvement, that such autonomous firms are probably the least likely to seek external funding since, as confirmed by Table 2.1, freedom is a major stimulus to formation, and clearly, a single founder business is the ultimate form of personal freedom. However, it is also clear from Table 2.2 that single entrepreneurship is a minority occurrence (i.e. in only 19% of cases), with multiple forms of entrepreneurship proving far more prevalent, in keeping with previous evidence (Oakey et al 1988).

TABLE 2.2 *Number of founders by growth in turnover (1987-91)*

	0-24%		25-100%		100+%		Total	
	N	%	N	%	N	%	N	%
One	2	9.5	8	23.5	10	20.4	20	19.2
Two	6	28.6	8	23.5	26	53.1	40	38.5
Three	6	28.6	6	17.6	12	24.5	24	23.1
Four or more	7	33.3	12	35.3	1	2.0	20	19.2
Total	21	100	34	100	49	100	104	100

Chi Squared = 21.6 P = 0.0014

Evidence on the theme of medium-term development is provided by Table 2.3 which indicates the current position of the founder in the firm. The most extreme change over time that can occur is the total loss of the founder to a survey firm. In 17 cases (i.e. 13%), the founder no longer had involvement with the firm concerned, and it was not surprising to note that in 11 (24%) of these cases, the firms concerned were those that had been acquired since formation. Table 2.3 clearly indicates that the absence of the founder is most prevalent in the biotechnology firms. While only 54% of this sectoral grouping retained their founder, the equivalent for electronics and software was respectively 78% and 85%. Chapter 3 will clearly indicate through a number of differing measures that the internal financial strength of biotechnology firms is the weakest of the three survey sectors. This reality consequently often involves external investors from birth in biotechnology firms, and such external involvement may either result in full acquisition (noted above) or a strong external say in internal management. In such circumstances where external involvement is strong, further financial support may be contingent on the removal of the original firm founder. Thus, there is likely to be a link between external financial dependence and the ejection of founding entrepreneurs. Indeed, this reality is a major reason, discussed throughout this book, why most firm founders would prefer, if at all possible, to remain financially independent.

TABLE 2.3 *Status of founder by sector*

	Biotechnology		Electronics		Software		Total	
	N	%	N	%	N	%	N	%
Owner/Managing Director/Chief Executive	23	53.5	35	77.8	40	85.1	98	72.6
Other executive post	5	11.6	4	8.9	4	8.5	13	9.6
Gone	11	25.6	3	6.7	3	6.4	17	12.6
Other	4	9.3	3	6.7			7	5.2
Total	43	100	45	100	47	100	135	100

Chi Squared = 16.09 P = 0.013

The sectoral origin of the main founder

The following information on the sectoral origin of the main founder of new high-technology businesses is of central importance to the main hypothesis of this chapter. It has been argued above that the previous activities of entrepreneurs have a strongly deterministic influence on the scope for new business formations. Put simply, a new firm founder, beginning a business in his "thirties", most often bases the new enterprise on the skills he has developed over the preceding ten to fifteen years. In a high-technology context, such skills may have been developed in either public (e.g. university; government research laboratory) or private sector employment (e.g. large industrial firm). Moreover, it might be further asserted that the propensity for either a public or private sector origin for new entrepreneurs might vary between industrial sectors, with more founders of industrial origin emanating from the electronics industry, while the public sector might be a more prevalent origin for new entrepreneurs in, say, the biotechnology industry. The higher level of expected public sector founders in the biotechnology industry partly stems from the embryonic nature of the technologies on which the new firms of this sector are founded, and the simple lack of well established existing large biotechnology firms from which new entrepreneurs might "spin off".

The survey on which this book is based provides a number of opportunities to test the above assertions on the sectoral origins of firm founders. Table 2.4 presents results to the most obvious question on origin by revealing the sectoral origin of the main founder's employment, immediately prior to formation. At a general level, the assertion that founders of new firms base them on previous work experience is supported by the observation that over half the respondent firms (i.e. 56%), were acknowledged to be in the same sector as the firm that provided employment to the main founder, immediately prior to the formation. In only 15% of cases was the main founder derived from a different sector to that of the new firm. Such phenomena may occur when, for example, a software engineer employed in an engineering firm decides to begin a new software firm, thus *technically* causing a sectoral change when, in fact, the individual is merely continuing with his personal specialism. Interestingly, the public sector was the source of new entrepreneurs in a further 22% of cases.

TABLE 2.4 *Origin of founder by sector*

	Biotechnology		Electronics		Software		Total	
	N	%	N	%	N	%	N	%
Same sector	15	34.9	36	81.8	23	50.0	74	55.6
Other sector	5	11.6	4	9.1	11	23.9	20	15.0
Public sector	21	48.8	1	2.3	7	15.2	29	21.8
Other	2	4.7	3	6.8	5	10.9	10	7.5
Total	43	100	44	100	46	100	133	100

Chi Squared = 36.9 P = 0.000

However, Table 2.4 also provides an insight into sectoral differences in the distribution of these data, that indicate strong significant statistical difference during Chi Squared testing at the $p = 0.00001\%$ level. Such sharp sectoral differences are contributed to by the 82% of electronics survey firms that "spun off" from other pre-existing electronics enterprises, while the equivalent figure for biotechnology and software was respectively 35% and 50%. As implied above, this strong result must, at least in part, stem from the relatively well established nature of the electronics industry when compared with the biotechnology and software sectors. The comparatively mature technologies of the electronics sector, together with the existence of a strong sectoral base that provides "spin off" potential and possible markets for new firm entrants to the industry, all assist the potential for new firm formation within this sector by facilitating new potential product niches for founders to fill, and by providing industrial markets into which these new products can be sold. Conversely, the biotechnology sector, for example, is comparatively new, where product technologies of new firms tend to be at the "basic" science end of the R&D continuum rather than "applied" or "development" oriented. Moreover, as noted above, few large established firms exist from which new entrepreneurs might "spin off", while there are also few industrial markets into which these new firm entrants might sell their industrial goods and/or services. It is therefore not surprising that another reason for the sectoral diversity in Table 2.4 is that almost half (i.e. 49%) of the biotechnology firm founders emanated from the public sector (e.g. higher education or government establishments), whereas the equivalent figures for electronics and software were respectively 2% and 15%.

However, in terms of the earlier argument that the origin of the founder has a strong influence on the type of firm he establishes, both the "same sector" and "public sector" results (which combined accounted for 77% of respondents in Table 2.4) tend to support the view that entrepreneurs' previous experiences largely determine the nature of their business at the time of formation. Most of the founders "spinning off" within the same sector as their previous industrial employer utilise technical skills they developed at their previous employment, while new firm founders from the public sector are mainly concerned with the development of technologies that were created when employed at a public sector institution (e.g. university or government laboratory).

TABLE 2.5 *Origin of main founder by region*

	Rest of Britain		South East and East Anglia		Total	
	N	%	N	%	N	%
Same sector	33	47.1	41	65.1	74	55.6
Other sector	8	11.4	12	19.0	20	15.0
Public sector	23	32.9	6	9.5	29	21.8
Other	6	8.6	4	6.3	10	7.5
Total	70	100	63	100	133	100

Chi Squared = 11.69 P = 0.00851

Geographical results on the sectoral origin of founders provide possible evidence of broadly spread agglomeration influences. Table 2.5 indicates a number of interesting results with regard to the origin of survey firm founders in different parts of Britain. It is probable that the high concentration of *all* high-technology sectors in the South East of Britain is a partial explanation for the significantly higher 65% level of "same sector" formations from the South East and East Anglia, when compared to the 47% level for the rest of Britain (Table 2.5). A major impact of any agglomeration has always been its ability to build on the advantages of an early concentration of industrial production through the support of later local "spin off" new firm formations. This has been noted to be particularly true for new firm formations in Silicon Valley (Rothwell and Zegveld 1982; Oakey 1984). Although the South Eastern quarter of England is a much larger area when compared with Silicon Valley, high-technology "spin offs" into the home counties around London from established intra-regional high-technology firms may be reflected in the evidence of Table 2.5.

However, the strong statistical difference in Table 2.5, revealed during Chi Squared testing to be significant at the $p = 0.008\%$ level, owes more to the sharp regional difference in terms of the public sector origin of new firm founders. While, in the rest of Britain, there was a 33% level of acknowledgement of this source, the equivalent figure for the South East and East Anglia was a much lower 10%. These results lend weight to the argument that, in areas where high-technology industry is not well developed, public sector institutions, notably in the form of regional universities, must play a much larger part in the development of new high-technology small firms within their local economies.

TABLE 2.6 *Previous employment location of main founder by region*

	Rest of Britain		South East and East Anglia		Total	
	N	%	N	%	N	%
Within 30 miles	47	65.3	39	63.9	86	64.7
Same region	6	8.3	13	21.3	19	14.3
Elsewhere in UK	17	23.6	3	4.9	20	15.0
Overseas	2	2.8	6	9.8	8	6.0
Total	72	100	61	100	133	100

Chi Squared = 14.31 P = 0.0025

The well established phenomenon of new firm founders beginning their new enterprises within the vicinity of their previous employment is generally confirmed in 65% of all survey cases, and in each region of Table 2.6. However, perhaps more interestingly, the "large area" agglomeration argument advanced above in terms of the South Eastern quarter of Britain is lent some further support by Table 2.6 through the presence of two features common in high-technology agglomerations. First, there is a higher level of *intra-regional* movement in the South East and East Anglia in 21% of cases, compared with 8% for the rest of Britain. This partly reflects the high concentration of high-technology industry in all parts of this South Eastern quarter (previously mentioned above). Second, there is a higher 10% level of international *in-migrations* to this South Eastern quarter, compared with 3% for the rest of Britain. The international attraction of high-technology specialist agglomerations as locations for internationally mobile high-technology entrepreneurs is also a well established phenomenon (Oakey et al 1988). Clearly, all things being equal, Cambridge is more likely to attract internationally mobile investment than Liverpool. However, this observation again reinforces the point made above on the importance of universities for the stimulation of high-technology development in regional economies. The higher 24% level of incidence of "elsewhere in the UK" as a main founder's previous employment location in the rest of Britain, when compared with the South East and East Anglia at 5%, must partly reflect the large size of the South East planning region, and the unwillingness of entrepreneurs, not previously employed in the South East, to suffer the increased costs of a South Eastern location when forming their new business. Both commercial and private housing costs can be prohibitive for any entrepreneur seeking to move into the South East from a cheaper more peripheral location, while South Eastern entrepreneurs "spinning off" from local incubator firms are accustomed to these high operating costs (see below for confirmation of this assertion). All these regional differences were sufficient to cause significant difference during Chi Squared testing to produce significance at the $p = 0.002\%$ level.

Links with the previous employer

The trauma of foundation in new high-technology firms can be reduced in certain circumstances if good relations are maintained with the founder's previous employer. Although, as discussed in Chapter 3 below, the "spin off" of new firms from a parent organisation may be an acrimonious event, it is also possible for the birth of such firms to take place with the full consent of the previous employer organisation. In such amicable cases, there is scope for considerable assistance by the "incubator" organisation in terms of technical assistance, accommodation, patronage or other minor forms of assistance. Indeed, in cases where the new firm is founded to develop a basic scientific discovery (e.g. the "spin off" of a new firm from a science laboratory of a university), the new formal organisational arrangement is unlikely to break strong personal links between the new firm founder (or founders) and university colleagues. Such links may also have a formal manifestation in that the university may hold an equity stake in the new firm.

TABLE 2.7 *Links with previous employer by ownership status of firm*

	Acquired		Independent		Total	
	N	%	N	%	N	%
Links	6	25.0	50	48.1	56	43.8
None	18	75.0	54	51.9	72	56.3
Total	24	100	104	100	128	100

Chi Squared = 4.22 P = 0.04

Given the potential benefits of a link with a previous employer, it is not surprising to note from Table 2.7 that almost half of the new firms in the survey (i.e. 44%) continued to maintain links with a previous employer. However, this table, also indicating the ownership status of survey firms, clearly reveals that, while 48% of independent firms maintained links with a previous employer, the equivalent figure for acquired firms was a much reduced 25%, producing significant difference during Chi Squared testing at the $p = 0.04\%$ level. This result must partly reflect the fact (established above) that the original founder is less likely to remain in a firm following acquisition, thus probably breaking the link with a previous employer. It is also likely that the availability of the larger resources of the acquiring firm largely negate the need for a link with the founder's previous employer.

TABLE 2.8 *Formality of links with previous employer by sector*

	Biotechnology		Electronics		Software		Total	
	N	%	N	%	N	%	N	%
Formal	17	85.0	8	42.1	10	55.6	35	61.4
Informal	2	10.0	6	31.6	8	28.1	16	28.1
Both	1	5.0	5	26.3			6	10.5
Total	20	100	19	100	18	100	57	100

Chi Squared = 14.17 P = 0.007

A subsequent question on whether the link was formal, informal or both, elicited the responses presented in Table 2.8. The striking result in this table is the observation that almost double the number of biotechnology firms claimed the link to be formal compared with the other two survey sectors. The 85% level of formal links with a previous employer is very high, judged by average industrial standards. However, as will be noted in many other parts of this book, the biotechnology sector is characterised by close knit information and material linkages between manufacturers and researchers. In many respects, the biotechnology industry is a research industry in which the traded goods are either R&D itself, or the materials that enable R&D to take place elsewhere. This observation is supported by Table 2.9 in which the nature of formal biotechnology links noted in Table 2.8 are explored. In the biotechnology case, these links are almost equally attributed to R&D contracting, sales and purchasing relationships with a previous employer. In the other two sectors, customer and supplier relationships predominated.

Premises

The provision of a satisfactory operating environment for a new high-technology enterprise contributes to the success of the new firm. Attractive premises of high quality are often important from a credibility viewpoint. While it might be acceptable for a jobbing engineering firm to be located in a dilapidated Victorian factory, new high-technology firms are generally expected to operate from sophisticated buildings. However, an often conflicting requirement of any new firm is that early operating costs must be kept to a minimum, implying that floor space costs should be as low as possible, although certain high-technology forms of production may require special features to be included in their premises. Thus, image, cost and special functions are all possible factors in the "premises equation".

TABLE 2.9 *Type of link with previous employer by sector*

	Biotechnology		Electronics		Software		Total	
	N	%	N	%	N	%	N	%
Research	4	20.0			2	12.5	6	11.3
Customer	5	25.0	5	29.4	6	38.5	16	30.2
Supplier	4	20.0	2	11.8			6	11.3
Customer and supplier	2	10.0	7	41.2			9	17.0
Other	5	25.0	3	17.6	8	50.0	16	30.2
Total	20	100	17	100	16	100	53	100

Chi Squared = 19.18 P = 0.14

TABLE 2.10 *Type of premises firm occupies by sector*

	Biotechnology		Electronics		Software		Total	
	N	%	N	%	N	%	N	%
Factory	14	31.8	33	71.7	1	2.1	48	35.0
Office	5	11.4	4	8.7	36	76.6	45	32.8
House	3	6.8			7	14.9	10	7.3
Factory/office	2	4.5	7	15.2	1	2.1	10	7.3
Labs plus	17	38.6					17	12.4
Office/workshop	2	4.5	2	4.3			4	2.9
Other	1	2.3			2	4.3	3	2.2
Total	44	100	46	100	47	100	137	100

Chi Squared = 127.2 P = 0.00001

A further complication in terms of the present investigation is the reality that the three sectors of the study are likely to require sharply differing types of premises, ranging from office space for many software firms, through laboratories for biotechnology producers, to traditional factories for firms engaged in electronics production. Table 2.10 confirms this potential diversity by indicating the type of premises survey firms occupy, by sector. While 72% of electronics firms occupied traditional factories or joint factory/office facilities, only 32% of biotechnology firms were located in factories, with a solitary software firm in such premises. Most software firms were located in offices which constituted 76% of their locations (often in inner city districts) while, as expected, 39% of biotechnology firms were located in laboratories. Interestingly, 15% of software firms continued to exist at the time of the survey in the private home of the founder. Clearly for clean, quiet, small scale operations, the personal home is the ultimate low cost location. However, since one of these firms was located in a council house, the high prestige requirement of a high-technology location was unlikely to have been met!

TABLE 2.11 *Area of premises (square feet) by sector*

Square feet	Biotechnology		Electronics		Software		Total	
	N	%	N	%	N	%	N	%
<1,000	4	9.5	2	4.5	12	25.5	18	13.5
1,000–9,999	20	47.6	24	54.5	31	66.0	75	56.4
10,000–19,999	9	21.4	11	25.0	3	6.4	23	17.3
20,000+>	9	21.4	7	15.9	1	2.1	17	12.8
Total	42	100	44	100	47	100	133	100

Chi Squared = 21.61 P = 0.0014

Evidence on the total area of factory floor space required by survey firms in Table 2.11 produced a number of interesting and, in some cases, predictable results that showed sectoral differences during Chi Squared testing, significant at the p = 0.001% level. First, the software firms had significantly lower floor space requirements than firms from either of the other two sectors, with 92% of their number requiring less than 10,000 square feet of floor space. A perhaps more surprising result from Table 2.11 is the similarity in requirements between electronics and biotechnology firms. The above noted factory orientation of electronics firms, and the possible subsequent need for larger premises is confirmed in Table 2.11, with 41% of these firms occupying premises of over 10,000 square feet in size. However, the almost identical space requirements of the biotechnology firms was more surprising, where 43% of these enterprises also required over 10,000 square feet of space. This result implies that the often research-based work performed by biotechnology firms must require substantial amounts of factory or laboratory floor space.

Interesting sectoral data is provided by Table 2.12 on the cost per square foot of premises within the survey firm sample. Perhaps not surprisingly, higher rents were paid by the biotechnology and software firms, indicating their significant respective reliance on generally more expensive office and laboratory facilities, compared with a factory bias for electronics firms. While 66% of electronics firms enjoyed rents of less than £5 per square foot, the equivalent figures for biotechnology and software producers were 39% and 33% respectively. Other regional data on the cost of premises is provided by Table 2.13 in which it is clear, as suggested above, that costs of a South East or East Anglian location are significantly higher than in the rest of Britain. While in the rest of Britain rents of less than £5 per square foot were enjoyed by 59% of survey firms, the equivalent figure for the South East and East Anglia was 29%, producing significant difference during Chi Squared testing at the p = 0.002% level. Generally, however, higher rents are only one source of additional costs for South Eastern firms and combine with labour and transport congestion costs to increase overall operating costs. None the less, it should be remembered that these premiums are probably worth paying for the other agglomeration *advantages* that a South Eastern location offers in terms of high labour skills, material supply choice and potential local customers.

TABLE 2.12 *Cost of premises per square foot by sector*

Square foot	Biotechnology		Electronics		Software		Total	
	N	%	N	%	N	%	N	%
< £5	14	38.9	23	65.7	14	32.6	51	44.7
£5-14	18	50.0	12	34.3	23	53.5	53	46.5
£15+>	4	11.1			6	14.0	10	8.8
Total	36	100	35	100	43	100	114	100

Chi Squared = 11.42 P = 0.022

TABLE 2.13 *Cost of premises per square foot by region*

Square foot	Rest of Britain		South East and East Anglia		Total	
	N	%	N	%	N	%
< £5	35	59.3	16	29.1	51	44.7
£5-£14	22	37.3	31	56.4	53	46.5
£15+>	2	3.4	8	14.5	10	8.8
Total	59	100	55	100	114	100

Chi Squared = 12.08 P = 0.0024

TABLE 2.14 *Incidence of special needs for premises by sector*

	Biotechnology		Electronics		Software		Total	
	N	%	N	%	N	%	N	%
Special needs	24	54.5	17	37.8	9	19.1	50	36.8
None	20	45.5	28	62.2	38	80.9	86	63.2
Total	44	100	45	100	47	100	136	100

Chi Squared = 12.27 P = 0.0022

A substantial minority 37% of survey firms claimed that their firm had special needs in terms of their premises, which was highest in the biotechnology sectoral sub-group at 55%, sufficient to produce a significant Chi Squared test at the p = 0.002% level (Table 2.14). In terms of the specific needs of firms, there was no predominant requirement. Needs ranged from special ventilation, through waste disposal to additional security. Overall, the general trend of these results suggests that survey firms have comparatively sophisticated premises needs that are likely to cause premises costs to be higher than the average for most other forms of manufacturing "start up".

Summary conclusions

The initial half of this chapter was concerned with the proposition that high-technology small firms founders may be strongly determined in their choice of new enterprises by their previous experience and expertise. It was also suggested that, in addition to being constrained to particular industrial sectors when seeking to establish a new high-technology enterprise, barriers to entry and growth might vary between different high-technology sectors. Subsequent detailed chapters of this book will attempt to further explore this assertion on the sectoral variability of entry and growth barriers. Indeed, Chapter 3 will explore critically important sectoral variations in financial parameters associated with funding the growth of newly established NTBFs.

However, this chapter has provided substantial evidence to confirm that the previous employment experience of the main founder plays a vital role in the formation of the new high-technology small firm, both through the pervasive importance of a "product idea" as a basis for a new venture and the evidence that there was a strong coincidence between the technology of the new firm and the technology of the previous industrial employer of the main founder. It was noted that, in the case of public sector "spin offs", the founder in many cases (i.e. academic entrepreneurship) was often determined in his choice of high-technology firm formation by his previous academic skills. Other more detailed functional investigations of the formation process revealed further ways in which the founding technology of the firm dictated the needs of the embryonic enterprise. For example, the initial determining of a software basis for a new firm through the development of computer science skills in a university lead to "downstream" functional requirements appropriate to this form of activity. The greater use of office facilities and their generally higher cost is one example of how functional requirements can be strongly influenced by the basic productive activities of a new high-technology small firm.

The remaining chapters of this book will be concerned with the possible ways in which the differing technologies adopted by new firms' founders influence their ability to survive and grow within sectors. If comprehensive information at this sectoral level can be collected to describe adequately the general problems of the included firms, it is likely that this level of aggregation will be the best vehicle for policy prescription. While it is impractical to design broad high-technology small firm policy to be tailored to the level of *each* individual firm, the fundamental premise of this book is that the current tendency to treat all high-technology small firms (or New Technology Based Firms [NTBFs]) as if they were a homogenous group is also unrealistic and counter-productive. The main task of the following chapters is to establish the varying requirements of different new high-technology small firms at individual sectoral levels in order that the future development of theories and policies might be pitched at a sectoral level, taking account of the differing needs and attributes of individual sectors as and when they occur.

CHAPTER 3

The Interacting Impacts of the Founding Product Technology and Funding on Firm Formation and Growth

This chapter represents an integrated consideration of the influence of basic product technology and funding that combine to shape the progress of the firms at "start up" and in the early years following formation. While a clear option in planning the structure of this book would have been to present evidence on basic firm product technology and finance in separate chapters, this amalgam is an attempt more realistically to present the interrelated impacts of basic *raison d'être* product technology and the necessary funding (internal and external) as factors that influence the strategic approach of NTBFs. Clearly, it is a major task of this key chapter of the book to test the assertion that the basic product technology of the new firm has a strong *determining* influence on the financial "room to manoeuvre" experienced by new firm founders in differing high-technology sectors. The chapter begins with a conceptual consideration of factors that influence formation strategy, followed by an empirical investigation of the origin and strategic development and funding of product technologies in the firm.

Product innovation behaviour, new firm formation and growth: some introductory conceptualisations

Regardless of whether a new firm has been formed to sell a new product, service or both, a critical early decision of the founder (or founders) is the point at which the firm is *formally* established. While the definition of the "date of birth" might appear simple, there are a number of key events, from the first conception of a product or service idea, to the achievement of initial sales, that *might* be termed the formation date. In particular, it is a risky exercise, when considering high-technology small firms, to define the formation point as "the date of incorporation" or the stage at which the firm "first began trading", since critical events in the product innovation process may precede these "milestones". The definition of formation used in this chapter is the year in which the founder first employed himself, or at least one worker, as a full-time employee of the new enterprise. While, to some extent, any definition is open to criticism, the full-time employment of staff is a precise and consistent point in the formation of the company. Although there has been considerable academic debate on the definition of the point of formation, since critically, it determines the *age* of a given firm (Mason 1983; Oakey et al 1988 etc.), it is important that a consistent definition be adopted, and that this definition be appropriate to a given methodology and subject of study. Judged in this manner, the above employment definition of the formation date, chosen for this book, performed well.

The decision to commit to the full-time operation of a business, in most cases, implies a sharp increase in investment by the founder (or founders). A particular problem with the formation of many high-technology small firms, which distinguishes them from other manufacturing and service sector enterprises, is the frequent lack of a marketable product, during and after formation, upon which early sales and the independent financial strength of the firm can be built. Indeed, as later evidence in this chapter will reveal, the ability of new firms to generate internal financial independence is strongly influenced by the sectoral origin of the firm *within* a general high-technology industrial classification. None the less, in cases where early sales *are* achieved, subsequent profits from these sales are often only a partial contribution to the financial needs of the firm, and are "swamped" by high "front end" R&D costs that exist in many high-technology sectors (e.g. biotechnology) (Freeman 1982; Oakey et al 1990). This reality, for many high-technology small firm founders, has a number of serious organisational ramifications in the early years after formation that, in their rush to become established, they may not fully realise. Such ramifications are postulated conceptually below. The generalised models of formation proposed will be useful in interpreting the results of the later empirical passages of this chapter.

Simplified models of new firm formation and growth

New high-technology firms may adopt one of *four* main variants in terms of an initial founding strategy:

Strategy one
The new firm begins life with a new product that has been fully developed and is ready for the market, thus yielding significant immediate sales with which to fund the further growth of the business.

Strategy two
The firm is formed *prior* to the full development of the product, but post-formation development is *very rapid* and substantial sales are quickly achieved (say within one year), sufficient to fund the future growth of the business.

Strategy three
The firm is founded prior to product development, the development period, set out in the founding business plan, is closely adhered to (possibly three to five years), and the firm eventually moves into profit on (or near) the predetermined date, providing an acceptable medium-term return on internal *and* external investment.

Strategy four

The firm is founded prior to product development, *but* the development period, established in the business plan, is not adhered to, and development continues beyond the expected product launch date. Further investment by existing (and perhaps new) holders of equity is required.

Clearly, movement from option one to four implies *increasing levels of risk* for both the founders, and where relevant, external lenders of investment capital. Further, problems associated with scenarios two to four could be avoided by *all* high-technology small firm "start-ups" using the "strategy one" method of establishment in which "front end" funding is minimised by the rapid achievement of sales and subsequent profits at or near formation. Indeed, the relevance of the above discussion on the point of formation is highly relevant here, since the decision concerning when to *formally* establish the firm might be seen as a strategic decision on when in a period of "gestation" it is most prudent to opt for formal "birth". None the less, this chapter will give further detailed consideration to options *two to four*, since these are the most frequently chosen strategies by high-technology firm founders, as part of the process of securing the investment needs of the firm.

The above strategic options two to four are discussed in more detail below, with the aid of Figures 3.1 to 3.3, which present the problems of "start up" and early growth in simplified graphical forms (based on an earlier diagram – see Oakey 1984, p. 94). In these models, profits are defined as sales of goods and services minus operating costs which, for the purposes of the current argument, do not include R&D expenditure, which is shown separately. The product concerned will not make an overall profit for the firm until the area of profit exceeds in size the area of R&D costs, over the product life cycle.

Strategy two (expanded)

Product launch soon after formation, rapidly growing sales *and* R&D cost reductions are achieved following formation (Figure 3.1).

Figure 3.1 New product development (strategy 2)

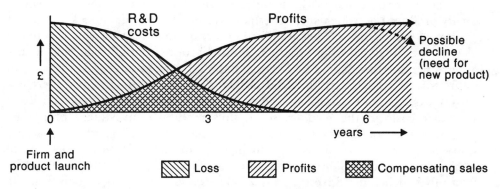

A "second best" option to formation with a fully developed product, this strategy of *rapid* product development has a number of advantages over the options depicted in

Figures 3.2 and 3.3 discussed below. Although the "front-end" funding of initial R&D is relatively costly compared to strategy one above, involving a degree of external assistance, it is short-lived, and the subsequent sales achieved have two benefits. First, for any external investors, there is rapid proof of product "saleability". Second, early compensating sales reduce the need for *external* support, ensuring that new firm founders have the *choice* of full control of the firm, or the involvement of external investors. In the following examples (i.e. Figures 3.2 to 3.3), it should be noted that, due to a lack of compensating sales following formation, such a choice is not an option. However, although product sales growth may be fast, product lives may be relatively short (see "possible decline" in Figure 3.1).

Strategy three (expanded)
Formation prior to product launch – launch deadline met (Figure 3.2).

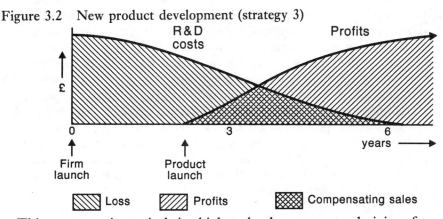

Figure 3.2 New product development (strategy 3)

This strategy is typical in high-technology sectors deriving from basic scientific discoveries in which the new product has received a degree of development in the public or private sectors in a university or incubator large firm, but further development is required before sales can be achieved (e.g. a new biotechnology enterprise). Substantial external investment is required to augment the internal capital of the founders during formation. A business plan is usually produced on which such external investment is based, and a key element of this plan will be the time between firm formation and the achievement of initial sales. As will be noted from Figure 3.2, initial profits from sales do not initially mitigate the, by then, falling R&D costs, but increasing sales suggest that overall profits will be achieved within the expected life of the product. Thus, there is a realistic expectation that total overall profit will substantially exceed total R&D and other costs in the medium term, thus rendering investment profitable. Increasing sales and growing profitability of the new firm will ensure that any equity held by external investors will increase in value. However, perhaps of equal value to external investors is the knowledge that self-sustained profits ensure that their initial investment *is all that is required,* and that such firms do not become "bottomless pits" into which further money must be invested in order to protect previous investment (Wilson 1993).

Strategy four (expanded)
Product development deadlines are exceeded (Figure 3.3).

Figure 3.3 New product development (strategy 4)

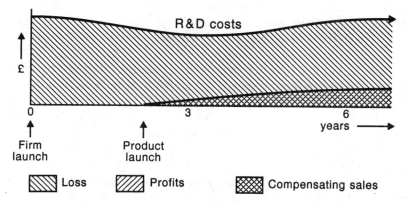

Problems associated with "near market" exploitation

Figure 3.3 conceptualises an extreme case in which, due to unforeseen problems with development, R&D costs continue to increase, and the product launch does not immediately prompt rapid sales and subsequent profit growth. Such a firm in which there has been considerable past investment, continues to be tantalisingly close to a major breakthrough which would redeem previous investment, but more investment is continually required to remove "final" technical barriers to large-scale market success, which persistently transpire to be penultimate! Clearly, any firm with no potential would not exist long enough to be a potential candidate for Figure 3.3. However, this model is common in parts of the biotechnology industry, where both the founders and external investors are often locked together in a stressful predicament in which only a *substantial* product success will allow the external investor an amicable "out". In extreme cases, new firms had not achieved *any* sales, more than five years after formation in the biotechnology sector (Oakey et al 1990). As previously noted, the easiest solution to all firm formation problems would be the strategic delay of firm formation until the conditions of formation strategy one above were met, in which the achievement of sales from a newly developed product *coincided* with the new firm launch. However, although many of the development costs of a new product would be removed from the new firm by adopting this strategy, it is clear that the cost of development would need to be borne by another institution, prior to establishment.

There are (at least) *two* potential explanations of why new firm founders in high-technology sectors may be inhibited from waiting until the proposed product is completely ready for market before launching a new firm:

1. The previous funder of the technical development does not foresee the potential of the proposed new product, or decides that such potential is not congruent with the aims of the organisation. Consequently the "incubator"

organisation refuses to support the project further. Such behaviour on the part of a university may occur when a development is considered "near market" applied research and therefore not appropriate for public sector support, which is mainly devoted to basic research. Moreover, the withdrawal of support has been noted as a major spur to new high-technology firm formation by technical entrepreneurs in the private sector when new small firms "splinter" from large "incubator" companies (Speigelman 1964; Farness 1968; Oakey 1981).

2. The new firm founder (or founders) wishes to acquire the intellectual property rights of the envisaged product through "spin off" at an early stage in the product's development. They fear that the development of a product idea to the "near market" stage would demonstrate the value of the technology to the legal owner of the embryonic intellectual property, who would then seek to maintain ownership. This phenomenon has been noted in research performed in the United States (Roberts 1991). Put simply, it is often *tactically* advisable to begin the new firm from an "appropriation of intellectual property" viewpoint, earlier than would be the case if internal financial strength was the sole criterion.

These assertions are useful in considering the recent debate on the role of public sector universities support for invention and innovation. As noted above, there is currently a misguided belief that the role of British universities is to perform *basic* scientific research. It is argued that, once a basic scientific idea is applied, the research becomes *near market*, and should be funded by the private sector. Apart from the reality that the development of a basic scientific discovery in an *applied* mode (e.g. lasers), frequently must also involve *further basic research* during application, there may be a funding gap caused by either the forced or voluntary removal of basic research from universities or large firm laboratories, *before* the private capital markets are willing to view this research as near market. For example, a university-developed technology may "spin off" in the form of a new firm, either because of its *public sector defined* near market applied status, or because of a desire by the founders to acquire intellectual property (stimuli noted above), five years prior to the envisaged market entry of the proposed product. However, private capital market investors may be prepared to invest in firms only two years before the commencement of sales are expected (usually in a business plan), thus causing a three-year "funding gap" (Figure 3.4).

It is hypothesised that Figure 3.4 represents a very real day-to-day problem for the founders of high-technology small firms. Indeed, a major reason why the scenario of over-optimistic lead times depicted in Figures 3.1 to 3.3 often occurs is that a firm founder constructing a business plan to attract external funding support may consciously (or subconsciously) reduce the lead time for the development of a five-year R&D programme (Figure 3.4) to an unrealistic two years in order to attract the required support. Although risky, this strategy has some merit in that a two-year programme *might* be achieved within budget, while the financial involvement of an external funder implies that further funding would be more likely, if necessary.

Figure 3.4 New product development

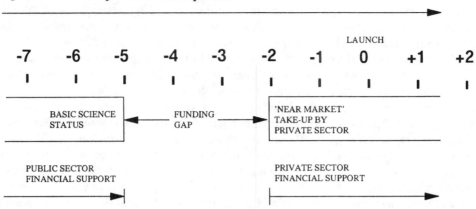

These initial ideas on the interaction between the founding product technology and the funding problems of new high-technology small firms are useful conceptual "benchmarks" to which the following evidence on how technological differences influence the formation and growth rates of firms within the study may be compared. The interaction between these hypothetical assertions and actual results will provide a useful point of departure for the conclusions of Chapter 7, where the formation and evolution of NTBFs will be re-examined. In the following empirically-based passages, initial consideration will be given to product strategies and sales performance, followed by consideration of more general financial criteria. Throughout this chapter, the implicit impacts of technological differences in these phenomena will be exposed through the examination of data from the three study sectors.

The derivation, mix and financial contribution of founding technologies

The origin of the founding product technology of the new firm

As implied above, the source of the founding product technology has an important bearing on point of firm formation, the rate of growth, and the subsequent *financial strength* of the firm in the post-formation period. For example, it might be argued that entrepreneurs forming a business on the basis of a new product with which they have "spun off" from the R&D departments of a university or large firm, would be more likely to be of the types shown in Figures 3.1 and 3.2 (implying substantial product development after formation), than would be the case for those founders beginning their businesses with a technologically *proven* product.

Table 3.1 contains initial evidence on the source of founding technology in survey firms. First, it is clear that the "technical entrepreneur" phenomenon is strong, since the largest proportion of 31% of firms were founded on the basis of the founders' technical knowledge. These results support the evidence in Chapter 2 on the sectoral origins of the firm founders. However, the derivation of a main product "after foundation" was almost of equal importance (i.e. 30%), while "acquired rights" (16%) and "common technology" (12%) were also represented. Table 3.1 also indicates notable evidence on the sectoral variations between the three study sectors. Overall, sharp differences are reflected in a Chi Squared test showing strong significance at the $p = 0.0007\%$ level. In particular, the strongest sectoral result was the high level of "main founder" origins in the biotechnology sub-group of firms, which was almost half of these establishments at 46%, almost double the percentage level of the other two sectors. A second more marginal variation in Table 3.1 is the higher proportion of software survey firms noting that the founding technology had been developed after formation (i.e. 38%). This difference is almost certainly explained by the rapid rate at which products can be conceived from a common technological base (i.e. various forms of computer language) and launched in the software industry, implying a short "lead time" between the formation of the firm and the product launch, and the often "bespoke" nature of the work performed. Thus, it should be noted that the above comments on funding development through long product development lead times (Figures 3.2 and 3.3) is less relevant in this sectoral case.

TABLE 3.1 *Origin of initiating technology by sector*

	Biotechnology		Electronics		Software		Total	
	N	%	N	%	N	%	N	%
Main founder	20	45.5	11	24.4	10	23.8	41	31.3
After foundation	11	25.0	12	26.7	16	38.1	39	29.8
Acquired rights	6	13.6	7	15.6	8	19.0	21	16.0
Common technology	3	6.8	13	28.9			16	12.2
Other	4	9.1	2	4.4	8	19.0	14	10.7
Total	44	100	45	100	42	100	131	100

Chi Squared = 27.02 P = 0.0007

A final sectoral feature of Table 3.1 is the high level of use of "common technology" in the electronics firms, where 29% of these enterprises acquired their founding technology in this manner, which was also 81% of *all* firms acknowledging a commonly available source. This result probably reflects the tendency for the electronics sector to possess a substantial sub-sector of hardware producers, where the technology involved is well established and copying is a popular follower technology strategy (e.g. printed circuit boards; transformers etc.).

The market launch of the founding technology, *and its contribution to the financial viability of the firm*

This sub-section builds upon the above consideration of product technology origins by dealing in detail with the specific contribution of the *founding technology* to the financial strength of the firm. In particular, the following analysis examines *whether* sales of the initiating product have yet been achieved, and if so, the year after formation in which the founding or "raison d'être" product or service achieved initial sales. Subsequently, details are given of the contribution to growth of this initiating technology, at the end of the first year after formation, and at the time of survey.

TABLE 3.2 *Proportion of total sales which initial technology comprised in first year of sales by unit profit margin*

	Unit profit margins							
% sales	< 50%		50-74%		75-100%		Total	
	N	%	N	%	N	%	N	%
< 50%	4	7.8	9	21.4	3	17.6	16	14.5
50-99%	10	19.6	8	19.0	8	47.1	26	23.6
100%	37	72.5	25	59.5	6	35.3	68	61.8
Total	51	100	42	100	17	100	110	100

Chi Squared = 10.65 P = 0.031

In terms of the survey results, all participant firm executives were asked if their firms had managed to obtain sales from their initiating technology. Given that any level of sales would elicit a "yes" response, and that a number of the biotechnology firms, which are probably the slowest to obtain substantial product sales, had gained some modest income from their initiating technology through related contract R&D services, the 96% overall level of acknowledgement of sales achievement is not surprising. Indeed, 92% of responding firms to a question on the date at which sales were first achieved stated that some revenue was produced by sales within the first year of full operation. Moreover, Table 3.2 clearly indicates that the initiating technology provided all of the value of sales in the first year of operation in a majority of cases where sales were achieved (i.e. 62%). While this observation is perhaps not surprising, Table 3.2 also indicates that there is a relationship between the propensity to derive all of corporate sales in the first year of operation from the initiating technology, and the unit profit margin produced by such a product or service. There is a statistically significant relationship between a high proportion of sales from their initiating technology in the first year and low profit margins on this technology (per unit of production). This phenomenon has two likely explanations. First, since it is lower technology firms that are more likely to achieve faster production and sales of the initiating technology, due to short "lead times" they may be more prevalent in the "100% from sales in the first year" grouping of Table 3.2. Second, a less attractive further common attribute of lower technology firms is their propensity to enjoy lower unit profit margins on their higher volume outputs (e.g. printed circuit boards). The converse is true for higher technology firms. While the eventual profit margins their products can command will be larger, the complexity of their development process means that they will be slower to bring their new technology to the market. It is likely that the spread of results in Table 3.2 reflects such differences.

Respondents were also asked to indicate the contribution to sales made by the initiating technology at the time of survey. Initially Table 3.3 confirms, as might be expected, that the proportion of firms claiming the initiating technology accounted for all of their sales declined sharply from 62% in the first year in Table 3.2 to 34% at the time of survey (Table 3.3). However, it is clear that, although the time of survey is constant for all survey firms, the duration of a product's contribution to the growth of the firm depends upon firm age. Thus, apart from the small number of cases where the initiating technology has become obsolete (discussed below), it is likely that the contribution of the initiating technology to the output of the firm will be influenced by the age of the firm. This is indeed the case in Table 3.3, where it is clear that the older survey firms tended to record a lower contribution from the initiating technology than their younger counterparts. For example, while 35% of firms of less than five years of age derived all of their sales from the initiating technology, the equivalent figure for firms over nine years old was 17%. This difference was sufficient to produce a Chi Squared test significant at the p = 0.005% level. It is probable that the trends in all of these data reflect the tendency in the high-technology sectors of the study for a generally rapid rate of technological change to ensure that constant product specification evolution is required to preserve a competitive edge. Indeed, "fear of obsolescence" *before* market entry can be achieved is a major concern for developers of sophisticated technologies where long lead times are involved. While the *eventual* rewards to such radically innovating enterprises *may* be great, the scope for a competitor to reach the market first with a better product is always a danger.

TABLE 3.3 *Proportion of total sales which initial technology comprised in 1991 by firm age*

% of sales from initial technology	0-5 Years		6-8 Years		9-13 Years		Total	
< 50%	13	32.5	25	69.4	29	53.7	49	37.7
50-99%	13	32.5	5	13.9	16	29.6	37	28.5
100%	14	35.0	6	16.7	9	16.7	44	33.8
Total	40	100	36	100	54	100	130	100

Chi Squared = 14.98 P = 0.005

The unit profit margins achieved by the initiating product or service technology

It is a common feature of manufacturing industry that the size of profit margins on products or services largely depend on the ubiquity or uniqueness of the technology embodied in the output. In the case of high-technology industry, this assertion can be seen in the observation that unit profit margins on printed circuit boards are low due to the ubiquitous nature of the necessary production technology which allows many producers to enter the market, and subsequently, through competition, force down the price of products. Conversely, profit margins on sophisticated technologies may be high due to their often unique performance specifications, consequent lack of competition, and scope for high prices (Oakey 1984). Moreover, it is also the case that items of high-technology production from small firms may occupy a pivotal strategic position in the product development programmes of larger customers.

For example, a critical software programme designed by a small software firm might ensure the success of the installation of several million pounds worth of manufacturing equipment. If an extreme case is imagined, a small software firm charging £50,000 for a software package that only cost £10,000 to produce is a price that will be willingly paid by a large patron if the product or service removes a critical technological bottleneck in an expensive programme of capital investment. It is also true, however, that the same large customer might haggle over pennies if a long-standing printed circuit board supplier increased prices by 5%. The main difference between the software and printed circuit board manufacturers is that the large firm customer knows that he can obtain the same or very similar quality of service in several local alternative printed circuit board suppliers, while the software programmer *may be* difficult or impossible to replace.

In terms of the three generally high-technology sectors considered in this study, it might be expected, in view of the assertions above, that the overall level of unit profit would be high when compared with the rest of manufacturing industry. In the next table, the percentage figure given denotes what is in common parlance often termed "mark up" or the difference between the total cost of production and the price charged to customers for a given unit of production. Thus, a 100% "mark up" would be a situation in which the unit sale price was twice its cost of production.

TABLE 3.4 *Unit profit margins by sector*

Unit profit margin	Biotechnology		Electronics		Software		Total	
	N	%	N	%	N	%	N	%
0-24%	11	31.4	6	15.0	4	11.1	21	18.9
25-49%	7	20.0	15	37.5	8	22.2	30	27.0
50-74%	10	28.6	18	45.0	14	38.9	42	37.8
75-100%	7	20.0	1	2.5	10	27.8	18	16.2
Total	35	100	40	100	36	100	111	100

Chi Squared = 16.21 P = 0.013

Table 3.4 presents an interesting breakdown of the margins of unit profit by the three sectors of the study. The first general observation worthy of note is that, as might be expected in high-technology sectors, there is a high level of unit profit margins, with 81% of the respondent firms enjoying unit profit margins of 25% or more. However, within this overall pattern there are a number of interesting sectoral differences. Taking the biotechnology sub-group of firms first, it is notable that these enterprises are comparatively evenly spread over the percentage profit margin categories. This even pattern of response probably reflects the varied range of activities within this generally heterogeneous sector. While the few product-based firms, often selling diagnostic kits or other bio-medical products, are likely to enjoy high profit margins on their output, equipment makers and R&D service consultants often have less scope for large "mark ups".

The electronics firms differ in that most fall in the middle two percentage categories between 25% and 75% (i.e. 83%), with only one respondent in the "over 75%" category. As mentioned above, this pattern of results probably reflects the observation that fewer electronics products are unique, and too high a profit margin charged on products will ensure that the customer merely sources from elsewhere. Finally, the majority of the software firms are in the higher percentage "mark up" categories, with the largest proportion of 28% of these firms in the over 75% grouping (Table 3.4). Overall, the software sector produces the largest proportion of firms in the "over 50%" grouping at 67%, compared to 48% for electronics and 49% for biotechnology. Although subtle, these differences between sectors combine to produce a Chi Squared test showing significant difference at the $p = 0.013\%$ level.

The marginally higher "mark ups" in the software case may be partly caused, not so much by the actual "mark up" margin above a moderate production cost, *but* the often very low cost of production *per se*. The apparent contradictory nature of such a statement raises a very subtle point which is fundamental to this chapter. Indeed, the percentage "mark up" is only partly a function of production costs. For example, it could be argued that if a software product cost £1,000 to produce, and it was company policy to "mark up" by 25%, the price would consequently be £1,250 to the customer. However, such a simplistic approach denies the clear truth that true genius is not measured in terms of "time spent" or material resources consumed. If, for example, the software programme produced for the cost of £100 through a sudden stroke of genius was unique and solved a previously insoluble problem, thus saving millions of pounds in a customer firm, the usual 25% "mark up" could be massively increased due to the product's negligible initial cost of production *and* high utility when applied in production, yet remain a bargain to the customer. Put simply, a key observation to be made from the above example is that the "intellectual value added" to outputs, which are the critical selling points of many high-technology products, are not directly related to the amount of human and financial resources invested in production (Oakey and Cooper 1989). In extreme cases, products in the electronics industry, which might cost hundreds of thousands of pounds to develop, manufacture and market, might return low unit profit margins, or even make a loss, while a software engineer *might* devise the basic concept for a world-beating software programme in less than an hour. In the software cases, noted in Table 3.4, the observed larger unit profit margins are likely to be a combination of very low costs of production and the high prices the high quality intellectual property can command. However, the above example has been exaggerated to make a valid point on the value of "intellectual value added" to the saleability of products. Later evidence on overall year end profitability suggests that high mark ups on individual products within the software firms are not generally reflected in the overall prosperity of the firms of this sector.

Changes in post-formation product mix

A major advantage of many new high-technology firms is that the "leading edge" nature of the new technology embodied in their initial products often acts as strong protection from competition in the initial years following formation. Unlike ubiquitously produced products, where the necessary production technology is widely known (and copied), "leading edge" technology-based new firms may be market leaders in key small niche areas of production, where product prices may be consequently high (Oakey 1984). Indeed, such an advantage can be prolonged if intellectual property protection can be obtained. However, a force that counteracts this privileged market position is the rapid advancement of the sectoral technology, of which such a product is part. If the computer sub-sector of the electronics industry is cited, a new technology-based firm manufacturing a small, yet powerful, transformer for the computer industry, may suddenly find that a major market will be lost if higher specification and smaller size requirements are not met as the large customer computer firm reduces the size of its product due to competition.

TABLE 3.5 *State of sales for initial product by sector*

	Biotechnology		Electronics		Software		Total	
	N	%	N	%	N	%	N	%
Growth	25	64.1	15	38.5	21	53.8	61	52.1
Stability	8	20.5	14	35.9	13	33.3	35	29.9
Decline	6	15.4	10	25.6	5	12.8	21	17.9
Total	39	100	39	100	39	100	117	100

Chi Squared = 6.27 P = 0.18

Thus, it is clear that new high-technology firms must be conceiving their second new product when the initial product is at the launch stage. It might also be expected that most of the new firms in this study, where 70% of the total have been founded for between five and thirteen years, will have introduced a further product to augment (or replace) their initial offering. Initially it was established, perhaps not surprisingly in such relatively new firms, that 86% of the 137 firm total were continuing to offer the product or service on which the firm was founded. However, the transient value of the initiating technology is confirmed by Table 3.5, where it is clear that in only 52% of cases was the initiating technology acknowledged as "still experiencing growth in sales", while in 30% of cases "stability" had occurred, and in a minority 18% of firms, decline had begun. While Table 3.5 is presented by sector, the only notable sign of sectoral difference is a slightly higher tendency for greater initial product decline in the electronics sub-group of firms (i.e. 26%), and a larger proportion of initial product growth firms in biotechnology (i.e. 64%). This is not surprising, given that rapid product life cycles and consequent obsolescence, discussed above, is generally most applicable (although not exclusive) to the electronics sector. These data on the contribution of the initiating technology to growth, suggest that such relatively new firms achieve high rates of product innovation change.

It is also implicit from Table 3.5, that a strong majority of firms must have introduced additional products since formation. In fact, 117 firms, 86% of the total, had introduced additional new products or services since foundation. The number of new products and/or services introduced is summarised in Table 3.6, where it is initially clear that, in the majority of cases (i.e. 59%), more than three products had been introduced, in addition to the initiating technology, since formation. Not surprisingly, increasing age was significantly associated with a propensity to introduce new products.

TABLE 3.6 *Number of major product lines and/or services on offer to customers by age*

Age	0-5 Years		6-8 Years		9-13 Years		Total	
Number of products	N	%	N	%	N	%	N	%
3	19	52.8	8	23.5	21	45.7	48	41.4
4-5	11	30.6	14	41.2	20	43.5	45	38.8
6+	6	16.7	12	35.3	5	10.9	23	19.8
Total	36	100	34	100	46	100	116	100

Chi Squared = 11.04 P = 0.026

TABLE 3.7 *Basis for future production of firm by sector*

	Biotechnology		Electronics		Software		Total	
	N	%	N	%	N	%	N	%
Original technology	23	53.5	19	41.3	31	67.4	73	54.1
New technology	15	34.9	22	47.8	14	30.4	51	37.8
Other	5	11.6	5	10.9	1	2.2	11	8.1
Total	43	100	46	100	46	100	135	100

Chi Squared = 7.95 P = 0.093

Areas of future production growth

This final theme on the relationship between product innovation strategy and the early life of NTBFs draws data from a survey question on whether the future production of the firm would rely on the existing product technology of the firm, or on new technology, yet to be exploited. Table 3.7 indicates that while marginally over half the respondents cited existing technology would provide future growth for the firm, a substantial minority of 38% of firms stated that the future lay in technology yet to be exploited. The substantial size of this minority group of firms testifies to the technological volatility of all the sectors included in this study. In terms of sectoral differences, although there are no significant statistical variations in Table 3.7, there is a notable difference between electronics and software in that software firms were more reliant on their original technology (i.e. 67%) when compared with their electronics counterparts (i.e. 41%). However, it is likely that this pattern of results merely reflects the strong core technology within software, where product variation is more commonly associated with function rather than the basic technological form.

The general funding of formation and growth

Clearly, a major component in the successful formation and subsequent growth of a high-technology small firm is adequate funding of the early life of the business. In this context the term "adequate" is critical because, as discussed in previous research on the funding of high-technology small firms (Oakey et al 1988, Chapter 9), the amount of investment required depends partly on the ambition of the founder (or founders). It is now well established that most new high-technology founders prefer to establish and grow their firms on the basis of a minimum of external (or borrowed) capital, while relying mainly on personal savings and/or early profits from sales (Oakey 1984; Deakins and Philpott 1994). This tendency was also explicit in the assertions of Figures 3.1 to 3.3 with which the findings of this sub-section on finance have a clear explanatory link. Indeed, Figures 3.1 to 3.3 suggest that the extent to which firms are able to generate early sales, influences the embryonic growth and independence of survey firms. These arguments have a direct impact on this consideration of firm funding, since the extent of external "front end" investment required by a new business must largely depend on an ability to reap compensating early profits from sales. Put simply, every extra pound that can be earned from early sales is a pound that does not need to be borrowed from external sources, thus saving the new firm money, and to bring the argument "full circle", allowing the founders to remain largely independent.

However, such simplistic assertions on the desire for financial independence, while generally sustainable, are complicated at the individual level of the firm. The extent of the need for external borrowing depends *partly* on the strategic aggression of the founder (or founders) of the new firm. In particular, risk-averse firm founders may choose to move forward slowly on the basis of retained profits, thus avoiding external funding involvement as argued above. However, it is clear from the arguments surrounding Figures 3.1 to 3.3, and evidence presented below, that the chosen product technology of the firm also *partly determines* the potential for financial independence. This tension between strategic approach and a variability in sectoral determinism provided by fluctuating early sales revenue potential between the new small firms of differing high-technology sectors is depicted in Figure 3.5, where the extent of external borrowing is anticipated in the light of possible growth strategies and general sectoral determination.

Figure 3.5 Hypothetical extent of external borrowing

	Electronics	Software	Biotechnology
Aggressive Growth	HIGH	HIGH	VERY HIGH
Risk Averse Growth	VERY LOW	VERY LOW	HIGH

Taking each sector in turn, and bearing in mind previous evidence on the contribution of early sales to independent financial growth between sectors (Oakey 1984; Oakey et al 1990), the following assertions are possible. In terms of electronics firms, as might be anticipated with any aggressive growth strategy, firms seeking to expand quickly will clearly need the assistance of external capital providers, while their risk-averse counterparts from within their sectors *can*, if they so wish, largely ignore the external assistance of financial institutions, and expand the business, in keeping with profit growth, thus maintaining a strong control on the management of the enterprise. Critically, the ability of firms in the electronics sector to produce early profits from sales may facilitate the potential for the *choice* of *either* aggressive *or* risk-averse strategies.

Similarly, the position of software producers, it is proposed in Figure 3.5, may be largely the same as that of electronics firms. A clear ability to develop very early sales revenue with a relatively low required investment in capital equipment, facilitates the choice of either an aggressive or risk-averse approach to growth. However, a peculiarity of the software industry is that, although barriers to entry in this high-technology sector may be low in comparison to other industries, subsequent growth after formation also may be low, certainly when compared to electronics. A plausible explanation for this particular pattern of formation and growth is that low barriers to entry in any sector is a phenomenon that implicitly carries a "down-side" characteristic of many entrants, subsequently causing a very competitive pricing structure for the sector, and sluggish growth after firm entry. Indeed, easy entry into software production by many producers may also mean that the finite total market size of the industry is, in the short term, a further constraint on small firm growth after formation.

However, this argument on competitive conditions within software production does not conflict with previous evidence on unit profit margins in survey firms which indicates high "mark ups" on software products. Unlike other producers in, for example, the electronics industry where multiple units of production are produced, software firms may only produce one unit of production in several weeks (i.e. a software programme). Thus a single substantial unit "mark up" for such a firm may represent the total profit for the firm for that period, whereas an electronics firm operating on a lower unit profit margin would achieve a total profit for the period by multiplying the "unit mark up" by thousands of units of output produced during the same period. Thus, when comparing the profitability of firms in different sectors over a common time period, a consideration of *both* unit profit margins of production and *total units produced* would need to be made. It is certainly true that the apparently high unit profit margins in the software industry have not resulted in evidence of year end profitability or firm growth considered later in this chapter. This is probably a reflection of the bespoke nature of much small firm production within this sector where, as a result, although unit profit margins on individual items of production are generous, large overall profits are inhibited by the low total number of units produced.

A consideration of the biotechnology sector in terms of Figure 3.5 probably provides the most striking observations. In particular, previous evidence (Oakey et al 1990) has indicated that, for most new biotechnology firms, the lead times on new product developments ensure that a sole reliance on internal financial resources is only a possibility for a small minority of new biotechnology firms (e.g. perhaps those relying mainly on R&D services to launch the new firm). As Figure 3.5 suggests, those firm founders seeking aggressive growth are likely to need a *very high* level of external financial gearing if the firm is to progress rapidly. Although such an approach might be possible, for example in the new bio-medical field where exciting new discoveries do occur that are attractive to external investors, this strategy is also highly risky. High and temporally intense levels of "front end" investment in new biotechnology enterprises may move the firm forward on an R&D front at an enhanced rate, but the current capital market suggests that high and intense investment must be followed by rapid and high returns if the support of essential external backers is to be sustained. Indeed, the "high" value given to "risk-averse" new members of the biotechnology sector in Figure 3.5 is an acknowledgement of the reality that even highly conservative founders within the biotechnology industry, who intuitively would prefer to remain financially independent in a similar manner to their counterparts in the electronics and software sectors, may be *forced* to obtain substantial external financial support, due to the inevitably long periods of development through which most biotechnology product development must travel en route to achieving sales as finished products. Evidence below on the extent of external investment in biotechnology firms at the time of formation, and the incidence of formal business plans in this sector, bear witness to an early realisation on the part of many founders that external financial help will be unavoidable when the firm is founded.

Although many of these assertions are based on previous empirical work performed by this author in other contexts, the summary conclusions at the end of this chapter will judge whether the arguments made above have been sustained by the evidence below. This financial analysis begins with a consideration of the extent to which survey firms relied on internal resources or external funds, both at the time of foundation, and subsequently.

Funding at the time of formation

The point of formation is a period of particular financial stress for the new firm. The cost of establishment involves unique expenses which occur simultaneously, including the purchase of accommodation, production equipment, and a host of other "once only" expenditures, which differ from the "running costs", that tend to be incremental (e.g. the purchasing of additional items of production equipment and labour). Additional problems with formation finance are caused by the frequent lack of any "track record" on which to base possible applications for external capital funding assistance. However, analysis begins with a consideration of personal funding.

Table 3.8 initially indicates the proportion of the formation capital which was derived from *personal* sources. Although the table shows no statistical significance in terms of major differences between sectors, three observations are prompted by these data. First, it is readily apparent that only 26% of respondent firms fell in the categories that implied a *mixture* of personal and other funding sources at the time of formation. However, the second striking feature of Table 3.8 is the substantial, and exactly equal, proportions of firms which either relied *solely* on personal funds, or contributed *nothing* to the cost of formation (i.e. both 37%)! This strongly dichotomised result must be partly explained by the frequently observed tendency among small firm founders to avoid external funding if at all possible (Oakey 1984; Oakey et al 1988). However, as discussed above in the introduction to this chapter, the period of existence between formation and self-sustainability through sales varies widely, and in some cases, may be so protracted that substantial external funding is unavoidable from the outset.

TABLE 3.8 *Formation capital from personal sources by firm sector*

% from personal sources	Biotechnology		Electronics		Software		Total	
	N	%	N	%	N	%	N	%
0%	18	42.9	16	37.2	14	31.1	48	36.9
1-39%	7	16.7	10	23.3	2	4.4	19	14.6
40-99%	4	9.5	4	9.3	7	15.6	15	11.5
100%	13	31.0	13	30.2	22	48.9	48	36.9
Total	42	100	43	100	45	100	130	100

Chi Squared = 9.95 P = 0.127

Interestingly, these arguments are given additional weight by the observation that 100% personal funding was most prevalent in the software sector (i.e. 49%) (Table 3.8). In software, where entry costs into production have been suggested to be lowest, own funds may suffice, but in biotechnology and electronics production external assistance is more frequently accepted as essential from the outset. The smaller number of firms in the "mixed" funding categories may be simply a function of the relative youth of many of the enterprises in this study (i.e. 82% of firm foundations after 1980, and 45% after 1985), since involvement with external organisations generally tends to increase in likelihood with age of enterprise. While there were no significant sectoral variations, as might be expected, a majority 77% of survey firms claimed that there had been no contribution from profits during their firm's first year of operation, while conversely a 15% minority of firms had grown to depend totally on profits by the end of their first year in business.

TABLE 3.9 *Contribution of external funding to the development process by sector*

% External funding	Biotechnology		Electronics		Software		Total	
	N	%	N	%	N	%	N	%
0%	19	45.2	22	52.4	34	75.6	75	58.1
1-50%	3	7.1	5	11.9	6	13.3	14	10.9
51-99%	8	19.0	8	19.0	2	4.4	18	14.0
100%	12	28.6	7	16.7	3	6.7	22	17.1
Total	42	100	42	100	45	100	129	100

Chi Squared = 15.22 P = 0.0186

Table 3.9 contains more detailed evidence on the incidence of external capital acquisition during the first year of operation. The significant difference at the $p = 0.018\%$ level produced by a Chi Squared test is caused by the smooth and gradual transformation in the importance of external funding as consideration moves from biotechnology, through electronics to software. For example, avoidance of external funding is lowest in biotechnology at 45%, compared with 52% for electronics and a much higher 76% for software. At the other extreme, total funding from external sources is highest in biotechnology at 29%, compared with 17% for electronics and 7% for software. Since, as noted above, profits in most survey firms during their first year, regardless of sectoral origin, are rarely large enough to be a major factor in funding the firm, the much larger external involvement in biotechnology firms is more likely a reflection in high start up costs. These costs involve both the numbers of initial workers employed, and initial capital equipment purchased, which is often for laboratory equipment rather than comparatively rapid profit realising production machinery. Conversely, the predominant reliance on internal funds in software firms during the first year of growth is likely to stem from rapidly realised profits and the low initial cost of establishment.

The contribution of external funding three years after formation

It has been argued earlier in this chapter that the main objective of most small firm founders is the achievement of the financial independence which stems from a strong increase in product or service sales. Thus, firms which rely on previously generated personal or external capital as the main means of funding a business three years after formation must begin to question the eventual ability of the enterprise to produce a return on investment. It might be expected, therefore, that the achievement in most firms of sales growth reduces the proportionate share of "own funds" to the financing of the firm as the firm becomes established and self-sustaining. Indeed, with regard to "personal investment" in the firm, while it will be remembered that a minority 37% of survey participant firms derived none of their investment from personal sources at the time of formation, this figure rose to a majority 84% three years after foundation, with each of the constituent three industrial sectors recording individual percentages in excess of 80% for firms with *no* personal funding.

However, the pattern of results for external funding of firms *after* three years indicates major sectoral differences in Table 3.10. There is a smooth and consistent change from biotechnology, through electronics to software in terms of the degree of external funding of individual firms after three years of existence, producing a Chi Squared value significant at the p = 0.002% level. For example, at the extremes, 29% of biotechnology firms relied totally on external funding three years after formation, compared with 8% for electronics and 4% for software. In terms of a total absence of external funding, 45% of biotechnology firms recorded this strongly independent position, compared with 74% of electronics and 85% of software firms. One clear implication from these results is that any "common" assumptions on the general growth (and return) potentials of high-technology firms in differing sectors is fraught with danger.

TABLE 3.10 *Contribution of external funding to the development process 3 years after formation by sector*

% External funding	Biotechnology		Electronics		Software		Total	
	N	%	N	%	N	%	N	%
0%	17	44.7	29	74.4	39	84.8	85	69.1
0–49%	4	10.5	2	5.1	4	8.7	10	8.1
50–99%	6	15.8	5	12.8	1	2.2	12	9.8
100%	11	28.9	3	7.7	2	4.3	16	13.0
Total	38	100	39	100	46	100	123	100

Chi Squared = 21.20 P = 0.002

The contribution of profits three years after formation

Although, to the novice, the distinction between profit on sales and overall year end profit might appear pedantic, it is perfectly possible in high-technology small firms to be making a substantial profit on the output it sells, but overall profitability is not achieved, either due to the low volume and subsequent value of (hopefully increasing) sales, or high R&D costs on further product development which exceed earnings from sales. Indeed, previous data has indicated the high level of multiple product development in these firms (see Table 3.6). Hopefully, it is safe to assume that firms with sales profit income, and no substantial external borrowing, will be solvent in the sense that income from sales is sufficient to meet the firm's needs. Conversely, since firms often borrow externally as a last resort, external borrowing is a reasonable indication that sales do not meet the full funding needs of the firm. It should be noted that this analysis excludes a minority of firms that had not reached three years of age by the time of survey.

Initial evidence of the fruits of a strong drive towards financial independence, based on strong sales performance, is revealed in Table 3.11 which again presents striking data on sectoral differences in the contribution of profits from sales to the total funding of the firm at the time of survey (i.e. net profit income from sales – *not* implying an *overall net year end* profit – year end profits will be considered below). A surprisingly large proportion of 57% of the survey firms were *totally* funded by sales profits within 3 years of foundation. However, a substantial minority 25% of these enterprises had *no* sales profits. Again there is a consistent and strong pattern of sectoral difference, with a strong contribution from sales profits among software firms, with 74% of businesses in this grouping obtaining all their funding from sales profits, compared to 58% of electronics firms and 34% of biotechnology enterprises. Indeed, exactly half of all biotechnology firms in the study received no profits from the sale of a product or service, compared with 7% for software. These sectoral variations produced extremely strong statistical difference from a Chi Squared test, significant at the p = 0.0002% level.

TABLE 3.11 *Contribution of internally generated profits to firm funding 3 years after formation by sector*

% Internal profits	Biotechnology		Electronics		Software		Total	
	N	%	N	%	N	%	N	%
0	19	50.0	9	22.5	3	6.5	31	25.0
1-99%	6	15.8	8	20.0	9	19.6	23	18.5
100%	13	34.2	23	57.5	34	73.9	70	56.5
Total	38	100	40	100	46	100	124	100

Chi Squared = 21.93 P = 0.0002

Perhaps the most striking feature of Table 3.11 is the clear evidence, which adds weight to earlier results that, at the extremes, an overall 82% of survey firms either relied *solely* on internal sales profits for their funding (i.e. 57%), or were totally dependent on external funding sources three years after foundation (i.e. 25%). Only 19% of survey firms were in a position to use sales revenue as a partial funder of the firm. In terms of the most extreme group of firms obtaining no capital from profits on sales, their numbers, as might be expected, declined from 102 in the first year of formation to 31 after three years. While 30 of these firms with no contribution from sales profits were derived from the 102 "start up" year group with no income from external sales, the remaining firm with no contribution from sales after three years was a firm gaining between 50 and 99% of investment from sales in the first year! However, it is significant to note that biotechnology firms comprised 19 (i.e. 61%) of the "die-hard" group of firms which had *never* received a contribution from sales profits. These data on the contribution of profits to the overall funding of the firm certainly suggest that the biotechnology sector possesses a disproportionate share of the firms which might be deemed synonymous with the example of Figure 3.3.

The involvement of external investors through equity shares

There is considerable evidence from previous research that, given a choice, founders only reluctantly seek the involvement of external investors (Oakey 1984; Oakey et al 1988; Mason and Harrison 1994). Therefore, in view of the above comments concerning Figure 3.5, where the constraining influence of commercial activity is stressed, it might be proposed that external financial support might be strongly related to the sectoral origin of new firms. In particular, it might be expected that more external control in the biotechnology sector firms might be apparent as a result of the unavoidable need for external help in the early years following formation, as new products are developed.

In terms of the degree of external equity involvement in survey firms at the time of formation, generally, it transpired that a minority 21% of the 131 responding firms acknowledged that external equity was initially held in their firms. In a series of various subsequent tests to ascertain if these externally involved firms showed any particular tendencies, the only analysis to prove significant was a comparison between sectors. Table 3.12 indicates a smoothly declining pattern of external equity involvement in survey firms, ranging from a 34% level in biotechnology enterprises, through 23% for electronics, and a very low 9% for software firms, causing significant statistical difference at the $p = 0.02\%$ level.

TABLE 3.12 *Externally held equity at formation by sector*

	Biotechnology		Electronics		Software		Total	
	N	%	N	%	N	%	N	%
External equity	14	34.1	10	22.7	4	8.7	28	21.4
None	27	65.9	34	77.3	42	91.3	103	78.6
Total	41	100	44	100	46	100	131	100

Chi Squared = 8.42 P = 0.015

Significantly, of the fourteen biotechnology firms with external ownership of equity, ten of these external investors were venture capital organisations. Indeed, these ten cases represented two-thirds of all the fifteen instances of venture capital involvement in all survey firms. It is also significant to note that, in general terms, the majority of 79% of firms with no external equity ownership at formation, bears witness to the above observation that most firms in all sectors are keen to achieve total internal independence if at all possible.

In order to gain an impression of how the pattern of external ownership of equity had evolved over the period since formation, firms were also asked to indicate the *current position* in terms of external equity involvement (Table 3.13). As might be expected, post-formation survival had involved an increase in overall external involvement which marginally increased from the 21% noted above to 29% by the time of survey. However, while total independence had been maintained in a majority of both electronics and software firms, independence in biotechnology firms had declined by 16% from 66% at formation to exactly half by the time of survey. This sharp increase in external involvement in the biotechnology sub-group of firms is reflected in the stronger significance of the Chi Squared test on Table 3.13, when compared to Table 3.12, which showed increased significance at the $p = 0.001$% level.

TABLE 3.13 *Externally held equity in 1991 by sector*

	Biotechnology		Electronics		Software		Total	
	N	%	N	%	N	%	N	%
External equity	20	50.0	12	26.1	6	13.6	38	29.2
None	20	50.0	34	73.9	38	86.4	92	70.8
Total	40	100	46	100	44	100	130	100

Chi Squared = 13.73 P = 0.001

Interestingly, the increase in external involvement in the biotechnology firms since formation was not provided by more venture capital firms, as might have been expected. While the level of venture capital firm investments remained unchanged at ten involvements, the main increase in external involvement in the biotechnology sub-group was an extra four private individuals who took equity stakes in survey firms. This phenomenon of "business angel" involvement might represent an embryonic trend, but also by default, may reflect a disenchantment on the part of venture capital firms with biotechnology new ventures in the late 1980s. Overall, the introductory assertions of this chapter that biotechnology firms would be more likely to be involved with external funders is supported by the general pattern of these results.

The degree to which external involvement in survey firms represents a real inhibition of managerial freedom is suggested by Table 3.14. Although numbers are small in this sub-group of firms with external equity ownership for which answers were provided, it is clear that in twelve of the twenty biotechnology firms the external holder of equity had gained a say in the management of the company. The sectoral diversity, especially between biotechnology and software, is borne out by the observation that, of the small minority of six software firms with external involvement, none of the external stakeholders exercised any management control. Clearly, the extent of "hands on" involvement by external holders of equity is partly a function of the size of the equity stake and the above noted involvement of venture capital firms in the biotechnology sub-group of survey firms must, to some extent, reflect the longevity of the involvement.

TABLE 3.14 *Role for shareholders in company management by sector*

	Biotechnology		Electronics		Software		Total	
	N	%	N	%	N	%	N	%
Role	12	66.7	3	27.3			15	42.9
None	6	33.3	8	72.7	6	100	20	57.1
Total	18	100	11	100	6	100	35	100

Chi Squared = 9.78 P = 0.008

Use of a formal business plan at the time of formation

Beyond the understandable motive of survival, it might be expected that relatively sophisticated NTBFs should have a growth strategy which, given favourable macro-economic conditions beyond the founder's control, the new firm should seek to adhere to as it expands its product or service activities. In the past, such motives might not gain any more expression than the private thoughts of founders, or through informal conversations with bank managers as a short-term overdraft might be negotiated. Indeed, many founders would view the formal writing down of their objectives as "incriminating evidence" which might be used against them at some future date if stated goals had not been achieved! However, with the increase of venture capital investment in high-technology small firms in the 1980s, and the clear requirement for the stating of objectives in a business plan as part of the bargain, there was a general "cultural shift" towards a climate where it was generally considered that all firms should construct a business plan, regardless of whether external financial assistance was sought or not. From the viewpoint of this book, apart from an interest in whether survey firms formulated a business plan at formation *per se*, the existence of a formal business plan is a valuable indirect measure of the general sophistication of study firms.

TABLE 3.15 *The existence of a business plan at formation by age*

	0-5 Years		6-8 Years		9-13 Years		Total	
	N	%	N	%	N	%	N	%
Business plan	36	87.8	28	70.0	33	63.5	97	72.9
None	5	12.2	12	30.0	19	36.5	36	27.1
Total	41	100	40	100	52	100	133	100

Chi Squared = 7.13 P = 0.028

The most interesting initial evidence to emerge from analysis of the behaviour of survey firms concerning the formulation of a business plan is provided in Table 3.15, where the above general comments on the growing tendency towards the construction of a business plan at the time of formation is confirmed by the decline of the proportion of firms claiming to have been founded on the basis of a business plan as age increases. While 88% of firms up to five years old had been founded with the aid of a formal business plan, the figure in the nine to thirteen year category was 64%, yielding significant difference during Chi Squared testing at the p = 0.028% level.

TABLE 3.16 *Business plan at the time of formation by sector*

	Biotechnology		Electronics		Software		Total	
	N	%	N	%	N	%	N	%
Business plan	40	90.9	32	72.7	27	57.4	99	73.3
None	4	9.1	12	27.3	20	42.6	36	26.7
Total	44	100	44	100	47	100	135	100

Chi Squared = 13.02 P = 0.001

In keeping with previous assertions, it might be argued that firms with an external equity stake would have a greater propensity to produce a business plan on which such a strategy might be based. Thus, it is no surprise to discover that the sharp sectoral differences indicated in the propensity to be involved in external equity arrangements, is also continued in this investigation of the presence or absence of a formal business plan in survey firms. Table 3.16 indicates strongly significant difference at the p = 0.001% level, which is caused by 91% of biotechnology firms beginning life on the basis of a business plan compared with a much lower 57% of software firms, with electronics firms following the pattern of earlier sectoral data by falling approximately half way between these two extremes at 73%. As discussed at many junctures throughout this book, the high propensity to use a business plan to found the firm must largely be a function of the variable need for substantial "front end" funding of the new firm to provide sustenance in advance of any expected revenue from sales. None the less, it is also clear that, at the other extreme, over half of the software firms, where total financial independence was very high, constructed a business plan at the time of formation. Clearly, in terms of establishing the credibility of an enterprise, the initial and most important group of individuals that need to be convinced are the original founders of the firm. Especially when more than one founder is involved, the formal stating of the objectives of a new enterprise is an important means of formalising an *agreed approach* through a business plan, regardless of whether or not it is subsequently used to obtain external capital.

TABLE 3.17 *Business plan used to get help by ownership status of firm*

	Acquired		Independent		Total	
	N	%	N	%	N	%
Business plan	15	83.3	45	57.7	60	62.5
None	3	16.7	33	42.3	36	37.5
Total	18	100	78	100	96	100

Chi Squared = 4.10 P = 0.042

A further interesting indicator that the formulation of a business plan is generally associated with a more outward looking approach to funding and growth, is provided by Table 3.17, in which a question on whether the business plan had been specifically intended to attract external capital investment is compared with whether the firm was subsequently acquired or not. A substantially higher 83% of firms that were subsequently acquired stated that the business plan had been formulated to obtain external financial assistance, compared with 58% of firms that remained totally independent. A persuasive argument to explain this trend has been provided in the introduction to this section where it was suggested that the propensity towards external financial involvement is strongly linked to the central ethos of the firm. If the objective of the firm is fast growth by whatever means, all forms of external involvement are acceptable (including full acquisition). However, other owners, who are clearly in a majority (Table 3.17), set a high value on independence as the firm's *raison d'être*. They will sacrifice growth for independence and expand, stagnate or decline on the basis of what the firm can earn from internal activities.

TABLE 3.18 *Business plan used to get help by percentage employment growth (1987-91)*

	0-24%		25%-100%		100+>%		Total	
	N	%	N	%	N	%	N	%
Business plan	14	46.7	18	66.7	23	76.7	55	63.2
None	16	53.3	9	33.3	7	23.3	32	36.8
Total	30	100	27	100	30	100	87	100

Chi Squared = 6.01 P = 0.05

The association of fast rates of growth, and a propensity for external involvement, is supported by Table 3.18 in which there is a clear relationship between those firms using their business plans to gain external financial help and greater employment growth. While 77% of firms using their business plan to attract external investment fell in the over 100% employment growth category, the figure for firms not adopting this approach was 23%, producing significant difference through a Chi Squared test at the p = 0.05% level. All the evidence of this section, and other work on high-technology small firms (Oakey 1984; Oakey et al 1988), suggests that firms wishing to preserve their independence will sacrifice growth for internal control and proceed on the basis of what they can internally produce as argued above. Whether this approach is an efficient strategy, however, should not be judged in the short term since slow growth over a long period may be more rewarding than unsustainably rapid growth (and possible closure or acquisition) over the short term.

A final interesting observation concerning the construction of a business plan in survey firms concerns whether the firms constructing a plan used external assistance in its formulation. Clearly, measurement of this tendency is also a general surrogate for the importance attached to the formulation of a plan. If the plan is only designed for internal reference, then the quality of its form and content is not so important. The general tendency for biotechnology firms to display greater efforts, both in terms of attracting external capital and formulating a business plan, is supported by Table 3.19, which clearly indicates that 63% of biotechnology firms sought external help with the formulation of the plan compared, at the other extreme, with only 25% of software firms, creating significant statistical difference as a result of Chi Squared testing at the p = 0.016% level. Moreover, the above assertion that eliciting external help with the business plan might be linked to an intention to use the document to obtain external help is convincingly confirmed by Table 3.20 which indicates that, while 91% of firms obtaining external help with the construction of the business plan used this document to apply for external assistance, a lower 47% of firms relying on internally derived plans subsequently applied for external financial help.

TABLE 3.19 *Help obtained to produce business plan by sector*

	Biotechnology		Electronics		Software		Total	
	N	%	N	%	N	%	N	%
Help	22	62.9	14	50.0	6	25.0	42	48.3
None	13	37.1	14	50.0	18	75.0	45	51.7
Total	35	100	28	100	24	100	87	100

Chi Squared = 8.22 P = 0.016

TABLE 3.20 *Incidence of business plan used to get external help*

	External help		None		Total	
	N	%	N	%	N	%
Use of business plan	38	90.5	21	46.7	28	67.8
None	4	9.5	24	53.3	28	32.2
Total	42	100	45	100	56	100

Chi Squared = 19.1 P = 0.00001

The current financial position of the firm

The contribution of profits from sales to the current *internal funding of the firm*
As discussed above in the section on unit profit margins (page 38), although firms might not be declaring an overall "year end" profit (discussed below), profits achieved through margins on individual sales may be contributing to the funding of the firm on a daily basis. Indeed, small firms may have several strategic reasons why year end profits are not declared, while there is a strong general desire among small firm managements to grow slowly on the basis of returns from sales. In such

circumstances, an optimal strategy for many high-technology small firm owners is expansion based on profits from sales in which the performance of the firm is approximately balanced at a "break even" point, which also has clear advantages from a corporation tax viewpoint. Thus, it should not be surprising to discover that this chapter records both a substantial contribution to the funding of the firm from sales profits (see page 48), while a substantial minority of firms examined on page 57 below in terms of year end profitability had not achieved an overall profit.

TABLE 3.21 *Contribution of "ploughed back" profits from sales by sector*

% Profits from sales	Biotechnology		Electronics		Software		Total	
	N	%	N	%	N	%	N	%
0%	9	22.0	2	4.8	6	14.3	17	13.6
1-49%	7	17.1	4	9.5			11	8.8
50-99%	6	14.6	10	23.8	4	9.5	20	16.0
100%	19	46.3	26	61.9	32	76.2	77	61.6
Total	41	100	42	100	42	100	125	100

Chi Squared = 17.19 P = 0.009

While page 48 above considered the contribution of sales profits to firm funding three years after formation, it is useful, in this final empirical section, to obtain a *current* view of this source of funding. The following analysis affords a common "benchmark" against which all survey firms may be judged since, as noted above, the data on funding after three years excluded a minority of enterprises that were less than three years old at the time of survey in 1991. Thus, the data presented below indicates the current position of all survey firms providing the required information. Table 3.21 indicates the extent to which "ploughed back" profits from sales currently contribute towards the general funding of the firm. While a comparison of the propensity to rely on sales profits with a range of potentially significant variables yielded no significant relationships, Table 3.21 clearly indicates that the exception to this "rule" is again the influence of sector. At the extremes, for example, biotechnology firms were the largest grouping in the less than 50% category with 39%, while only 46% of firms in this industry relied totally on profits for the funding of the firm, compared with 62% for electronics and 76% for software companies. Overall, the 62% of total survey sample firms relying on "ploughed back profits" for all of their funding is conclusive proof that a conservative reliance on internal funding is the only strategy entertained by a majority of these NTBF owners. However, Table 3.21 also supports other evidence in this study by suggesting that this drive for independence is an option that may be constrained by differing high-technology industrial sectoral characteristics.

Evidence on the overall year end profitability of the firm in 1991

In ideal circumstances, evidence on overall profitability levels would be the most suitable measure of success for new high-technology small firms. However, in reality, information on profits must be digested with the utmost care for two major reasons. First, respondents to interview surveys of new firms may be reluctant to divulge any information on profits, and in cases where information is provided, it may be often deliberately misleading. Second, in many cases, firm managements have the option to arrange their accounts to not show a profit through manipulation of directors' salary levels and other "costs" which erode gross profits. Moreover, in cases where a firm has been acquired, there is scope for using overall corporate financial features to obscure the performance on individual subsidiary plants.

Conversely, the registering of an operating profit is a clear means of indicating to actual and potential external investors that, at the very least, the firm concerned is "going in the right direction". Indeed, it should be remembered that the declaration of an operating profit may not negate the need for external assistance in the NTBFs of this study. In circumstances where large amounts of "front end" funding over a number of years is required, substantial external funds may need to be added to growing internal profits as the firm seeks to develop the full potential of its product technologies.

TABLE 3.22   *Declared profits for 1990-91 by sector*

	Biotechnology		Electronics		Software		Total	
	N	%	N	%	N	%	N	%
Profit	15	36.6	30	66.7	30	65.2	75	56.8
None	26	63.4	15	33.3	16	34.8	57	43.2
Total	41	100	45	100	46	100	132	100

Chi Squared = 9.94 $P = 0.007$

Notwithstanding the above reservations on profit data, the achievement of year end profits may be an important growth "milestone" in the life of the firm (although it clearly should not be assumed that such an achievement is irrevocable)! Table 3.22 provides data on responses to a simple question on whether the firms had declared profits in the financial year 1990-91. A substantial minority of 43% of firms had not achieved this goal. However, in sectoral terms, while over 60% of both the electronics and software firms had achieved profits, the figure for biotechnology was a much lower 37%, causing a Chi Squared test to be strongly significant at the $p = 0.007\%$ level. This poorer performance by biotechnology firms supports widespread earlier evidence in this chapter which suggests that biotechnology firms are particularly slow in developing self-sustaining funding.

More detailed data compares the level of profits with the sectoral origin of firms in Table 3.23. While it must be remembered from Table 3.22 that only a small proportion of biotechnology firms declared a profit, Table 3.23 clearly indicates that, for firms acknowledging a year end profit, a significantly larger proportion of software firms fell into the less than £50,000 profit category (i.e. 66%) when compared to the other sectors. Indeed, 67% of electronics firms fell in the over £50,000 category. This is the first empirical confirmation that, although the software sector is relatively easy to enter, subsequent growth may be slow. The possibility that the lower profit levels in the software industry might be due to smaller firm size is confirmed by the observation that, while 77% of software firms employed less than 20 workers, the equivalent figure for electronics and biotechnology was 28% and 50% respectively. The possibility that software firms might have been smaller due to their younger age, however, was disproved by surprisingly similar sectoral age ranges. Indeed, it might be argued that the small size of software firms is conversely determined by their inability to generate profits. As might be expected, however, there was a general relationship between age of firm and the level of profits declared. Table 3.24 reveals that while 70% of firms of less than five years of age fell in the less than £50,000 profit category, 67% of firms of over 9 years of age fell in the over £50,000 grouping, causing sufficient difference to produce a Chi Squared test, significant at the p = 0.032% level.

TABLE 3.23   *Extent of year end profits by sector*

£	Biotechnology		Electronics		Software		Total	
	N	%	N	%	N	%	N	%
Under 50k	5	50.0	9	33.3	19	65.5	33	50.0
Over 50k	5	50.0	18	66.7	10	34.5	33	50.0
Total	10	100	27	100	29	100	66	100

Chi Squared = 5.80 P = 0.055

TABLE 3.24   *Extent of year end profits by age*

£	0-5 Years		6-8 Years		9-13 Years		Total	
	N	%	N	%	N	%	N	%
Under 50k	16	69.6	6	50.0	10	33.3	32	49.2
Over 50k	7	30.4	6	50.0	20	66.7	33	50.8
Total	23	100	12	100	30	100	65	100

Chi Squared = 6.84 P = 0.032

The extent of additional external funding for specific projects in established firms

Clearly, the ability of certain survey firms to produce profits, combined with a high degree of independence sought by many individual firm managements, will both respectively influence the need *and* inclination of firm founders to seek external capital assistance. However, even the most independently minded firm management may be attracted to external investment capital if it is available under favourable terms and conditions, as a means of augmenting internal resources.

Earlier sections of this chapter have previously dealt with external equity involvement, both at the time of formation, and currently. However, a question was put to managers of firms that had existed more than three years to ask if they had sought any *additional* external capital investment, more than three years after formation, to fund a *specific new project* (i.e. *beyond* any permanent funding arrangements in the firm, such as a founding equity stake). In keeping with much of the evidence of this chapter, sector was the variable that produced significant difference in the survey sample. Table 3.25 supports earlier evidence by clearly showing that there is a smooth trend in terms of such external involvement to fund a specific project, from a high 74% in biotechnology firms, through 48% in the electronics case, to 29% for software producers.

TABLE 3.25 *The existence of external funding after 3 years by sector*

	Biotechnology		Electronics		Software		Total	
	N	%	N	%	N	%	N	%
Funding	29	74.4	21	47.7	12	29.3	62	50.0
None	10	25.6	23	52.3	29	70.7	62	50.0
Total	39	100	44	100	41	100	124	100

Chi Squared = 16.40 P = 0.003

The combination of the need for extensive front end funding in the biotechnology sector exposed earlier in this book at a general investment level, must also partly explain the above result. While firms in other sectors might be more financially introspective, and better able to fund new projects from internal financial resources, biotechnology firms are often forced to familiarise themselves with external means of financial support at an early stage, and develop both the "culture" and expertise to seek external support for new projects. Table 3.26 also supports earlier evidence by indicating that it is the faster growing firms that are more likely to seek external financial support for the enterprises. Only 33% of firms with less than 25% employment growth sought external financial help, compared with 60% of respondents recording over 100% employment increases, producing a Chi Squared test significant at the p = 0.036% level. It is probable that this trend is partly explained by the observation that it is the more enterprising firms that are both most likely to experience employment growth and be aware of opportunities for external funding of specific projects, where possible.

Interestingly, further investigation of the main source of this external funding conclusively revealed that government was the principal provider of such additional project-specific funding. Indeed, in Table 3.27, 80% of the funding sources were attributable to the government, with banks a poor second at 13%. The popularity of government funding as a main source of financial support for particular projects is probably related to the project specific nature of much of the aid available (e.g. The

Enterprise Initiative; SMART awards), and the often free or subsidised nature of such assistance. In some circumstances, projects may be tailored or created to suit existing schemes, or dormant projects acted upon when government funding becomes available. The comparatively onerous conditions imposed by banks may explain their poor showing in Table 3.27, while the medium-term complexity of setting up venture capital involvement would only be appropriate for particularly large investment projects, reflected in the low incidence of this source in Table 3.27. Only 18 of the survey total of firms (i.e. 13%) claimed that they had been refused external assistance to develop specific projects. However, this pattern of response is more likely a reflection of the unwillingness of respondent firms to become involved with external providers of capital, than of their need for such funding. This assertion is partially reflected in the popularity of government funding, where terms are usually more attractive than the private sector, and government involvement does not normally threaten the independent survival of the small firm concerned.

TABLE 3.26 *The existence of external funding after 3 years by employment growth (1987-91)*

	Low to 25%		25% to 100%		100% and over		Total	
	N	%	N	%	N	%	N	%
Funding	13	33.3	21	56.8	24	60.0	58	50.0
None	26	66.7	16	43.2	16	40.0	58	50.0
Total	39	100	37	100	40	100	116	100

Chi Squared = 6.61 P = 0.037

TABLE 3.27 *Source of external funding by sector*

Source	Biotechnology		Electronics		Software		Total	
	N	%	N	%	N	%	N	%
Bank	1	3.3	4	19.0	3	23.1	8	12.5
Government	26	86.7	16	76.2	9	69.2	51	79.7
Venture capital	1	3.3	1	4.8			2	3.1
Other	2	6.7			1	7.7	3	4.7
Total	30	100	21	100	13	100	64	100

Chi Squared = 6.37 P = 0.38

The funding of future growth

In terms of future funding within the firm, Table 3.28 clearly shows that projected profits were perceived as a major source of funding in most firms. Indeed, 86% of survey firms suggested that over 50% of their future financial needs would be provided by anticipated profits. As noted above, biotechnology firms were marginally less likely to cite profits as a major future funding source, with a lower 60% of firms in the over 75% category compared to an 82% level for electronics and 81% for software. This result further contributes to a consistent pattern of evidence which suggests that the "lead time" for biotechnology products is substantially longer than for other high-technology sectors.

TABLE 3.28 *Expected proportionate contribution of profits to the total investment required by sector 1991-96*

% contribution	Biotechnology		Electronics		Software		Total	
	N	%	N	%	N	%	N	%
0-24%	4	16.0	1	2.9	2	5.4	7	7.3
25-49%	3	12.0	2	5.9	1	2.7	6	6.3
50-74%	3	12.0	3	8.8	4	10.8	10	10.4
75-100%	15	60.0	28	82.4	30	81.1	73	76.0
Total	25	100	34	100	37	100	96	100

Chi Squared = 7.034 P = 0.318

Summary conclusions

The general impression created by the data in this chapter strongly suggests that investment in new high-technology small firms is not a means of obtaining a fast return on investment. Although the recent deep recession may have exacerbated the effect, the majority of firms not recording a year end profit in 1990-91, and the widespread evidence of extensive and continuing external financial support of survey firms, were clear indications that high-technology manufacturing in NTBFs is an activity fraught with operational difficulties. Moreover, the expected sectoral differences in internal financial strength were readily noticeable in much of the evidence presented above, where the move towards profitability and self-sustained growth was particularly slow in the biotechnology firms of this study. Probably the best overall performance was detected in the electronics firms, where the movement towards self-sustainability was slower than the software firms, but faster than biotechnology enterprises. None the less, the slower *longer-term* growth of software firms when compared with their electronics counterparts, after profitability was achieved, suggested that electronics firms may have the best overall performance. However, the current position is merely a "snapshot" of performances at the time of survey. It may transpire that, in the long term (say twenty years), it *might* be individual biotechnology firms that prove to be the most successful – if they can develop viable product technologies and sustain funding over the difficult early years.

These comments on development time-scales lead to a consideration of external support funding for new high-technology small firms. The introductory comments of this chapter on the importance of profits from sales in providing independence for the firm, and a possible funding gap, are both concepts for which there is evidence in the above data. A number of survey firms experienced the extreme circumstances depicted in Figure 3.3, where sales profits from the initiating technology had not reached a point where they had exceeded overall operating costs. Moreover, it is also probable that, in keeping with these introductory arguments, a number of survey firms displayed over-optimism regarding product development lead times, which may have been forced upon them by the short-termist attitude of external funders (discussed in conjunction with Figure 3.4). Thus, it might well be the case that firms recording no year end profits after five years are performing optimally in an objective sense, but are two years "behind" when judged against an initially unrealistic business plan. This unhealthy set of circumstances is mainly caused by the business plans of high-technology small firms being designed to meet the short-term earning requirements of funding organisations (e.g. banks, venture capital houses), rather than the medium-term survival needs of the firms to which investment capital is extended. Evidence to support the view that a longer-term approach to high-technology small firm funding should be adopted was provided by data indicating a relationship between increasing age of firms and higher profit levels. However, the best levels of profit only appeared in firms of nine years of age or older. Waiting for nine years to gain any substantial return on investment would not be an attractive investment scenario for many current venture capitalists, who would prefer to seek an "out" well before the ninth year.

In such difficult circumstances, it is not surprising to find that a substantial minority of firms in this survey continue to completely avoid external financial involvement. This should not be interpreted as an indication of the lack of any need for external financial help, but can be more realistically interpreted as a result of an unwillingness to incur the onerous and unrealistic targets and conditions that would be required to attract external investment. Indeed, in a final question to all survey respondents on the *overall* barriers to future growth, 51% of the survey population directly acknowledged a shortage of finance as the main bottleneck to growth. These firms were a cross-section of survey firms indicating both strong independence and external involvement.

A final summation of the results of this chapter suggests that general conditions in the three sectors surveyed are not conducive to innovation and growth, with extreme difficulties experienced in biotechnology firms. In particular, an inability in a substantial minority of firms to develop year end profits, or a substantial contribution from this source, has produced two major funding problems. First, the obvious lack of internal financial independence means that external assistance, and its attendant uncertainty, remains a *necessity* to ensure survival. Second, this lack of profits, often after a substantial period of existence, ensures that gaining continued or additional external financial support will be increasingly difficult. In one sense the firms experiencing growth difficulties in this study represent less extreme examples since they continue to exist, where firms representing more severe cases have closed. Evidence in this chapter on the *continuing* importance of product innovation in these generally new firms suggest that the funding of R&D is a not only a problem for "start up" enterprises, but continues throughout the young life of the firm. Given these circumstances, the medium to long-term funding requirements of such high-technology enterprises will continue to be demanding.

Although barriers to entry may be highest in the biotechnology sector, as suggested by much of the evidence above, and lowest in software production, the greatest long-term profits *may* be provided by biotechnology firms in general, and bio-medical firms in particular. However, the "seeing through" of such firms to a sustained and successful future might involve a commitment in the ten to fifteen year range by banks and venture capitalists, rather than in terms of the three to five years common in present circumstances. In current conditions, it is likely that firms with this long-term potential will either close, due to a lack of continued financial support, or be absorbed into large firms in the same or related sectors due to acquisition (an outcome discussed in Chapter 6). The lack of any substantial numbers of new firms in many of the survey sectors in the medium sized (500 to 1,000) range bears witness to an absence of growth from small to large status. Much of the evidence of this chapter suggests that this absence may well be due to a lack of substantial *long-term* funding for high-technology small firms with promise. The evidence and considerations emanating from this chapter will play a major role in shaping the conclusions to this book in Chapter 7.

CHAPTER 4

The Role of R&D in High-Technology Small Firm Formation and Growth

Introduction

As previously discussed in Chapter 3 above, most new independent firms must invest capital at the time of formation in order to create, develop or perfect the products and services on which their early growth is to be based (Freeman 1982; Rothwell and Zegveld 1982). However, it is also clear from the evidence of Chapter 3 that, although the firms of this survey are *generally* of a high-technology nature, early investment capital has not been employed in similar amounts. Indeed, if a *broader* comparative view of the founding technological needs of a wide range of formation technologies is taken, the manner in which capital is invested in R&D during formation, and the speed with which returns are gained, can be argued to differ *sharply*, at the extremes, between high- and low-technology-based firms. A major reason for much higher levels of R&D investment and longer investment time horizons in NTBFs, when compared with their low-technology counterparts, is that they are frequently required to invest capital in parts of the innovation process that can be either neglected or completely negated by low-technology (often sub-contract oriented) enterprises, of comparable age and size. Moreover, these additional areas of NTBF investment cause enhanced danger since, not only are they an *additional* requirement, but they are also *particularly risky* since the, often necessary, extended length of "pay-back" period involved does not *assure* a good eventual return on investment.

If for the purposes of illustrating this argument extreme cases are proposed, a new biotechnology firm of the type common in this study is compared with a new sub-contract mechanical engineering enterprise in Figure 4.1. This model argues that, not only does the high-technology biotechnology firm operate a much longer innovation cycle, comprising additional stages, ranging from the commencement of R&D to final marketing, but also the engineering firm has a completely *different focus* on formation investment, with a much higher concentration on sophisticated production equipment, which, conversely, is of less importance to its biotechnology counterpart. Indeed, it has been observed previously that, rather ironically, low-technology firms often acquire expensive high-technology process machinery (e.g.

CNC lathes; work centres; robots), while much high-technology production in NTBFs tends to be labour intensive, in *both* R&D and production (Oakey 1981; 1984). However, the overall impression given by Figure 4.1 is that the production of a product or saleable service in the biotechnology instance is much more expensive, due to additional innovation functions performed during the innovation process.

In the case of the sub-contract engineering firm, assuming buoyant demand for its services, capital out-laid on a new process machine can be recouped through sales of the output produced by the application of this machine in production. In such circumstances, there is a clear and quantifiable link between the investment and the return over a relatively short time-span. Moreover, in terms of the total set of functions within the innovation cycle depicted in Figure 4.1, overall operating costs for this sub-contract firm frequently are reduced by the tendering system in which the work performed is specified by the customer, negating the need for product development through R&D, while the successful winning of a contract negates the need for expensive marketing. The complete converse is the case for the biotechnology firm. The critical point of relevance to this chapter is that, even in instances where the R&D does result in a highly lucrative product innovation in terms of eventual sales for the biotechnology firm, the period between investment and acceptable return will be far longer than in the low-technology example of Figure 4.1.

Figure 4.1 Model of high- and low-technology product development showing degrees of commitment

	R&D	Prototype development	Process Innovation	Marketing
New Subcontract Mechanical Engineering Firm	NIL	NIL	HIGH	LOW
New Biotechnology Firm	HIGH	HIGH	LOW	HIGH

TIME

However, the *very existence* of new biotechnology firms that are successful in attracting external capital support for their formation and growth suggests that the above generally negative comments on costs and risks must be offset by compensating attractions. Novel product ideas, forming the *raison d'être* for new firms, may range from "basic research" based ideas at one extreme, to variations on well established technological themes, and even crude copying, at the more "development research" oriented end of the research continuum. To the novice observer, it might appear that an optimal strategic approach for any firm founder would be *only* to form firms when a product technology has reached the development stage, thus reducing the required level of investment in "front end" R&D and subsequent risk (also discussed in Chapter 3 above). However, there are two major reasons why such an approach is often not possible in an R&D context. First, although it is generally true that the cost of initial investment in R&D reduces as technological development moves from basic to development oriented R&D commitment, it is generally also true that the potential competitive value of R&D output reduces commensurately. In particular, a relatively low R&D investment in well-established areas of technology will not lead to the exclusive intellectual property that will deliver market power and high prices. Minor variations on a well-established theme will only produce marginal advantages in markets where ubiquitously available technology ensures that there will be many market entrants and subsequent competitors, and lower prices.

Second, perhaps of more importance in determining the point of formation, is the environment from which the founder (or founders) emanated prior to the formation of the new firm. Generally it might be argued that firm founders from institutions concerned with basic science (e.g. commercial R&D laboratories or universities) will emerge to form firms based on more basic technologies than founders from established industries who learned their technological skills in the more "bread and butter" areas of technology at the development end of the continuum. In this general sense, firm founders base the technologies of the new firm they found on what they know, and in the short to medium term, this tends to determine the strategic options of such founders in terms of founding technological choice. Importantly, it thus follows that, for example, a lecturer in a biology department of a university intending to form a new firm based on his research will tend to be research oriented, implying the need for substantial "front end" R&D funding and risk if the product is to be developed. In this sense, there are few strategic options for this firm founder, and his "choice" of founding technology is determined by the nature of the idea that usually results from his previous work experience over the preceding ten to fifteen years.

Clearly, a major attraction inherent in new high-technology small firms is their potential for abnormally high profits from sales *if* protracted product development through R&D is particularly successful in terms of the eventual sales of the marketed product. The general hypothesised theme of determinism, adopted throughout this book, is again appropriate to this consideration of R&D investment in that the technological nature of the founding product will largely determine the amount of "front end" R&D required in order to bring an envisaged innovation to the point at which sales can be achieved. Indeed, in such circumstances, where the founding product of the firm critically dictates a strategy that is dominated by R&D investment needs, variations in management quality can have only a marginal impact on success (as argued in Chapter 2).

Since it has been argued in Chapter 3 that most firm founders would prefer to remain financially independent if possible, the results of this chapter on R&D investment may be seen as a more detailed causal explanation of the financial and strategic behaviour presented in Chapter 3. Put simply, because R&D is a major component of early formation and growth costs in most NTBFs when judged in terms of *all* industrial technologies, the extent of R&D performed in survey firms observed below has a direct bearing on the degree to which "front end" investment is required in survey firms after formation, and importantly from a managerial control angle, the degree to which total financial and managerial independence can be pursued within individual firms. Thus, it might be expected that a number of the sectoral trends for technological strategy and financial control noted in Chapter 3 will be reinforced by the empirical evidence of R&D investment in this chapter which is likely to be a major strategic cause of these financial effects. Indeed, issues concerning the impact of product development "lead times" on formation strategy will be further discussed in the light of *all* the results from this book in Chapter 7.

The extent of R&D commitment in survey firms

Although the above assertions on wide sectoral differences in R&D commitment are reasonable when pitched at an industry-wide level, the first observation to arise from Table 4.1 on the existence of an "in-house" R&D investment within *this* study of NTBFs is the very high overall R&D commitment across the high-technology study sectors, with 88% of firms acknowledging "in-house" R&D activity. This over 80% commitment testifies to the pervasive importance of R&D in providing a competitive edge for the small firms in these differing sectoral markets. However, as noted in previous work on high-technology small firms (Oakey 1984), the evidence of Table 4.1 is a rather crude measure of the existence of R&D. Later detailed information on the extent of this commitment in the firms acknowledging such an activity might show more sectoral diversity. None the less, this "presence or absence of R&D" variable initially was tested against a range of other variables in order to record any sign of significance. Interestingly, the only significant

TABLE 4.1 *In-house R&D by sector*

	Biotechnology		Electronics		Software		Total	
	N	%	N	%	N	%	N	%
In-house R&D	37	84.1	41	89.1	42	89.4	120	87.6
None	7	15.9	5	10.9	5	10.6	17	12.4
Total	44	100	46	100	47	100	137	100

Chi Squared = 0.73 P = 0.69

relationship to emerge in Table 4.2 was a strong association between the presence of R&D and high unit profit margins in survey firms. Put another way, the appearance of twelve of the total thirteen firms with no R&D capacity in the less than 50% category, almost certainly reflects the reality, noted throughout this book, that firms with more established low-technology products are less likely to maintain an R&D capacity, and that the products of these firms, due to their technological ubiquity, will not be able to command the high unit profit margins enjoyed by their more high-technology-oriented counterparts. However, such a "deficiency" should not necessarily be judged a disadvantage since, as discussed in Chapter 3, overall financial success depends not only on unit profit margins, but also on the number of units produced. Since low-technology firms are more likely to produce a higher volume of output, it should not necessarily be assumed that they are less successful in overall terms.

Table 4.3 reports the results to a more detailed question on whether the R&D acknowledged by the 120 firms in Table 4.1 was of a full- or part-time nature. A majority 62% of the firms with an R&D capacity claimed that this was full-time in nature (i.e. employing at least one full-time member of staff). However, in this case, Table 4.3. indicates a strong relationship between the maintenance of full-time R&D staff and strong employment growth. In particular, while only 26% of firms with only part-time R&D fell in the over 100% employment growth category, the percentage for firms with a full-time R&D commitment was 69%. This difference was sufficient to produce a Chi Squared test significant at the p = 0.0078% level. This is clear evidence that firms with a substantial commitment to R&D experience strong employment growth, although such expansion might not be a measure of commercial success for all firms if, as suggested by the following evidence, a proportion of these additional workers are R&D staff who do not contribute to the output of the new firm in the short term.

TABLE 4.2 *In-house R&D by unit profit margins*

	0-49%		50-100%		Total	
	N	%	N	%	N	%
In-house R&D	39	76.5	59	98.3	98	88.3
None	12	23.5	1	1.7	13	11.7
Total	51	100	60	100	111	100

Chi Squared = 12.74 P = 0.0004

TABLE 4.3 *R&D on a full- or part-time basis by percentage employment growth (1987-91)*

	Under 25%		25-100%		100+>%		Total	
	N	%	N	%	N	%	N	%
Part	21	58.3	11	30.6	10	25.6	42	37.8
Full	15	41.7	25	69.4	29	74.4	69	62.2
Total	36	100	36	100	39	100	111	100

Chi Squared = 9.71 P = 0.0078

TABLE 4.4 *Number of R&D workers by sector*

No. of workers	Biotechnology		Electronics		Software		Total	
	N	%	N	%	N	%	N	%
1-4	7	25.9	13	50.0	13	59.1	33	44.0
5-9	8	29.6	7	26.9	5	22.7	20	26.7
10-19	5	18.5	6	23.1	3	13.6	14	18.7
20+	7	25.9			1	4.5	8	10.7
Total	27	100	26	100	22	100	75	100

Chi Squared = 13.6 P = 0.035

Indeed, more detailed evidence on the numbers of R&D workers employed in survey firms is provided in Table 4.4. This tabulation provides the first evidence of sharp differences in the extent of R&D commitment in survey firms across the survey sectors. The most divergent sector, in keeping with much of the evidence of Chapter 3, is biotechnology, with almost half the proportion of firms in the one to four worker category at 26%, when compared with electronics (50%) and software (59%). Moreover, a perhaps more surprising feature of Table 4.4 is the significant minority of seven biotechnology firms that employed over twenty workers in their R&D departments. This is a strong surrogate for a very high level of R&D spend in these biotechnology firms and must explain much of the funding problems of new firms in this sector, for it is unlikely that this is a level of R&D investment that can be funded from profits or personal savings by new firms in their early years. Thus, such intensive "front end" R&D investment necessarily implies a high level of external funding of such enterprises.

A further interesting variable to show a significant relationship with numbers of R&D workers employed was the propensity for survey firms to have been acquired. Table 4.5 clearly indicates that there is a tendency for firms with large numbers of R&D workers to have become acquired. In particular, while 22% of independent firms fell into the 10 workers or more category, the equivalent figure for acquired firms was 56%, a pattern of results that yielded a Chi Squared test significant at the p = 0.008% level. It is likely that this relationship has been caused by the above noted financial stress that is put on firms with high levels of "front end" R&D requirements that creates the need for additional external financial support. Clearly, acquisition is one means of solving this funding problem (discussed in Chapter 6).

TABLE 4.5 *Number of R&D workers by ownership status of firm*

No. of workers	Acquired		Independent		Total	
	N	%	N	%	N	%
<10 workers	7	43.8	46	78.0	53	70.7
10 workers +>	9	56.3	13	22.0	22	29.3
Total	16	100	59	100	75	100

Chi Squared = 7.109 P = 0.0077

This final sub-section under a consideration of the extent of R&D commitment in survey firms considers direct estimates of the cost of R&D work at the time of formation, and at the time of survey. Initially striking evidence is provided by Table 4.6 on expenditure on R&D in the founding year by sector. Again, biotechnology firms differ from their counterparts in other sectors, with over half of their number appearing in the greater than £50,000 category, compared with 14% for electronics and 11% for software, sufficient to produce a Chi Squared test significant at the p = 0.0005% level. This represents a continuation of exceptional data on R&D in the biotechnology sub-group of firms. Such a high level of R&D spend in the formation year is further *clear* evidence of the strain that the cost of R&D expenditure puts on new biotechnology firms, in this case giving an indication of, not only the sums involved, but also the urgency with which such capital is utilised. Clearly, this is one of the reasons why "ploughed back" profits are not a significant factor in the funding of new biotechnology firms since there is precious little time for such compensating profits to accrue if, as Table 4.6 confirms, R&D expenditure is both high *and early* in the life of the firm.

TABLE 4.6 *R&D expenditure in formation year by sector*

R&D spend (£)	Biotechnology		Electronics		Software		Total	
	N	%	N	%	N	%	N	%
Under 50k	10	45.5	25	86.2	24	88.9	59	75.6
Over 50k	12	54.5	4	13.8	3	11.1	19	24.4
Total	22	100	29	100	27	100	78	100

Chi Squared = 15.21 P = 0.0005

Further interesting evidence is provided by Table 4.7 on the level of R&D funding at the time of survey. Initially it should be acknowledged that if a particular sector contained firms that were disproportionately young, this might distort the following sectoral results on R&D spend at the time of survey, since newer firms are generally more likely to spend less on R&D. However, there was no discernible age biases in the individual survey sectors that might distort the following evidence on R&D spending by sector. Following from Table 4.6, Table 4.7 reflects a predictable increase in R&D spending as firms become established, with the overall 76% of firms in the less than £50,000 spend category in Table 4.6 at formation declining to 37% at the time of survey (Table 4.7). However, the most striking feature of Table 4.7, beyond the general observation that a substantial 63% of the survey total spent £50,000 or more per annum on R&D, is the again surprisingly high level of R&D spend of at least £500,000 per year in a minority of biotechnology firms, a trend suggested by the employment figures of Table 4.4, and confirmed here in Table 4.7. These data, together with other evidence from this sub-section on R&D commitment, begin to indicate, not only the existence of an extremely high R&D funding requirement in most survey firms, but *also* the particular intensity of this funding need in a minority of biotechnology enterprises.

TABLE 4.7 *Sector by R&D expenditure in 1991*

R&D expenditure	Biotechnology		Electronics		Software		Total	
	N	%	N	%	N	%	N	%
< 50,000	7	24.1	13	37.1	15	48.4	35	36.8
50,000–499,999	14	48.3	22	62.9	16	51.6	52	54.7
500,000 + >	8	27.6					8	8.4
Total	29	100	35	100	31	100	95	100

Chi Squared = 21.30 P = 0.00028

Research and Development linkages

Although the above evidence on the extent of internal R&D within survey firms supports existing wide-ranging evidence that internal R&D is the main means by which technical advantage is developed in high-technology small firms (Freeman 1982; Oakey 1984), the technical specialisms of such enterprises often prompt incoming and outgoing technical information sharing links within external collaborators. Technical information networking, often as a support to internal R&D effort, is an efficient means by which engineering problems may be solved in a two-way process in which external technical assistance may be sought to remove an internal problem, or consultancy advice may be afforded to an external R&D partner. As observed in previous research (Oakey 1984; Oakey et al 1990), the technical experts within high-technology small firms may be "world leaders" in their own specialised "niche production" area. In such circumstances, notwithstanding intellectual property constraints, a community of experts may interact across firm boundaries in the private sector, and exist between private sector researchers and their public sector counterparts, notably within universities. New firms in the biotechnology sector have displayed many of these technical linkage characteristics (Oakey et al 1990). Thus, given the generally high-technology nature of the firms examined in this book, further consideration of the extent of R&D linkages is warranted below.

R&D sub-contracting for external agencies

Given the above noted clear high level technical expertise present in many survey firms as represented by their internal R&D commitment, it might be anticipated that these enterprises would perform work for other organisations. Indeed, it has been noted previously in this book that the performance of contract R&D services for large firm patrons is a relatively low risk method of beginning a new firm in that the early capital of the firm can be rapidly earned from R&D contract research activities which subsequently act as a financial basis for new product development as the firm becomes established. This approach has been noted to be particularly prevalent in the biotechnology sector (Oakey et al 1990).

TABLE 4.8 *Incidence of contract R&D by sector*

	Biotechnology		Electronics		Software		Total	
	N	%	N	%	N	%	N	%
Contract R&D	17	45.9	16	37.2	9	21.4	42	34.4
None	20	54.1	27	62.8	33	78.6	80	65.6
Total	37	100	43	100	42	100	122	100

Chi Squared = 5.47 P = 0.065

Such an observation on the biotechnology industry is supported by the evidence of Table 4.8, in which almost half (46%) of the biotechnology sub-sample performed contract R&D for an external organisation. The marginal failure of Table 4.8 to show significant difference (i.e. significant at the p = 0.06% level), is mainly caused by significant minorities of firms performing external consultancy in all the survey sectors (i.e. averaging 34%). However, beyond this presence or absence measure, Table 4.9 clearly indicates that, when survey sectors are examined to measure the value of contract R&D as a proportion of total R&D spending, a clearly higher 64% of biotechnology firms acknowledged that over 50% of their R&D spend was devoted to contract R&D, compared with a 36% level for electronics and software combined, recording statistical difference at the p = 0.04% level. The importance of contract R&D to the biotechnology firms may partly explain the high level of R&D investment noted for this sector above, since workers engaged on contract R&D work ensure that more of the risk of failure is borne by the customer, and generally tend to "pay their keep" more readily than is the case of "in house" speculative R&D work performed on an internal new product development. None the less, this "contract research effect" must only be partial since none of the biotechnology firms were exclusively engaged on contract work and all were producing at least one "own product".

TABLE 4.9 *Sectoral groupings by contract R&D as percentage of total R&D expenditure*

Contract R & D as % of total	Biotechnology		Electronics and software		Total	
	N	%	N	%	N	%
<50%	8	47.1	18	78.3	26	65.0
50%+>	9	52.9	5	21.7	14	35.0
Total	17	100	23	100	40	100

Chi Squared = 4.18 P = 0.04

Research and Development contracts awarded to external agencies

As part of an *interacting* R&D networking relationship it might be expected, as suggested above, that in certain circumstances, survey firms would elicit technical support through the "putting out" of R&D work to relevant external sub-contractors. The wide international R&D linkage relationships of the biotechnology survey firms (also evident in terms of material linkages in Chapter 5), is confirmed in Table 4.10 by their dominance of a smaller number of firms contracting out R&D, when compared with firms that performed contract R&D work (i.e. in Table 4.9). Although not statistically significant, 34% of biotechnology firms contracted out R&D work, compared with 24% for electronics and 15% for software. The greater tendency for R&D networking in biotechnology firms must be partly explained by the generally more "basic" nature of research in this sector which often continues to benefit in the private sector from an information sharing ethos more common in public sector universities, with which many of the survey firm's founders maintain informal personal contacts.

TABLE 4.10 *Sub-contracted R&D by sector*

	Biotechnology		Electronics		Software		Total	
	N	%	N	%	N	%	N	%
Sub-contracted R&D	15	34.1	11	23.9	7	14.9	33	24.1
None	29	65.9	35	76.1	40	85.1	47	85.1
Total	44	100	46	100	47	100	80	100

Chi Squared = 4.59 P = 0.0102

TABLE 4.11 *Important source of technical information by sector*

	Biotechnology		Electronics		Software		Total	
	N	%	N	%	N	%	N	%
Important source	32	72.7	29	64.4	19	40.4	80	58.8
None	12	27.3	16	35.6	28	59.6	56	41.2
Total	44	100	45	100	47	100	136	100

Chi Squared = 10.67 P = 0.00483

Important sources of external technical information

Small firms may also make use of a range of occasional technical information sources in order to back-up internal innovative efforts of the firm. Typically, universities or research associations fall into this category of supportive external technical information sources, important in the development of new product technologies. Thus, survey firms were asked if they maintained an external technical information contact important to the development of products and/or services in their firm. Again, Table 4.11 clearly indicates that the previously more outward looking attitude of biotechnology firms is maintained in answer to this question. A common general pattern of results has been replicated in which biotechnology firms represent one extreme with a 73% level of acknowledgement of an important external technical link, with electronics at 64% and software at the low extreme of 40%. This spread of results must, to a large extent, reflect the extent to which firms in the various sectors are able to be self-contained in terms of the product or service innovations they produce. Since it is clear from the above evidence that a majority of all firms perform significant R&D, the high level of external involvement by biotechnology firms must reflect the complex nature of their R&D activities while, conversely, software firms, relying mainly on "in house" software skills, appear more able to remain self-sufficient in terms of the extent to which they access external technical assistance.

TABLE 4.12 *Important source of technical information by age*

	0-5 Years		6-8 Years		9-13 Years		Total	
	N	%	N	%	N	%	N	%
Important source	28	68.3	27	67.5	23	43.4	78	58.2
None	13	31.7	13	32.5	30	56.6	56	41.8
Total	41	100	40	100	53	100	134	100

Chi Squared = 7.91 P = 0.019

Another interesting result to emerge from wide-ranging analysis was the observation that there was a surprising relationship between declining use of important external technical information link and increasing age. Table 4.12 clearly shows that, while 68% of firms less than five years old maintained important external technical links, the proportion for firms in the nine years and over category was a much reduced 43%, a difference significant at the $p = 0.019\%$ level. This pattern of results suggests that there may be a tendency towards introspection as the firm ages and an internal culture of "best practice" emerges. However, whether this tendency towards internal reliance is a rational interpretation of a reduced value from external input to internal R&D over time, or represents an irrational form of laziness, is hard to judge.

Valuable contextual insights are provided by Table 4.13, which indicates the *type* of external technical contact maintained by survey firms. As might be expected, the highest level of external links with universities was held by the biotechnology group of firms at 73%, while the other strong result from Table 4.13 was produced by the 68% of software firms that maintained strong external link with industrial firms. The diverse origins of external technical contacts displayed by biotechnology and software firms further supports the growing impression that biotechnology firms are more oriented towards basic scientific expertise, while software firms, due to the nature of much of their work in writing software to solve industrial problems, tend to be more practically oriented in terms of their external technical contacts.

TABLE 4.13 *Important source details by sector*

	Biotechnology		Electronics		Software		Total	
	N	%	N	%	N	%	N	%
Universities/ Academia	22	73.3	13	46.4	3	15.8	38	49.4
Libraries/ Literature	3	10.0	4	14.3	2	10.5	9	11.7
Industrial firms			4	14.3	13	68.4	17	22.1
Professional bodies	3	10.0	4	14.3			7	9.1
Government research departments	2	6.7	1	3.6	1	5.3	4	5.2
Other			2	7.1			2	2.6
Total	30	100	28	100	19	100	77	100

Chi Squared = 40.37 P = 0.00001

Further insight is given on the information in Table 4.13 by Table 4.14 which presents the results to a question enquiring if the external source of technical information was of particular importance to the firm's internal R&D effort. Although not recording any statistically significant difference, it is notable that a much larger 75% of the software firms acknowledged their external technical link to be of practical importance to their internal R&D effort. Thus, taking all these data on the software sector together, it would appear that although their overall level of external contacts was lower when compared with the other two sectors, the contacts that did exist tended to be of particular importance to the internal R&D effort of the firm. The strong industrial orientation of these links, indicated in Table 4.13, suggests that the strength of the contacts derives from the need to design the software to the task to meet a specific customer need, involving critical contextual information that can only be provided by the customer.

TABLE 4.14 *Technical information source important to R&D by sector*

	Biotechnology		Electronics		Software		Total	
Important	17	53.1	14	53.8	15	75.0	46	59.0
Not important	15	46.9	12	46.2	5	25.0	32	41.0
Total	32	100	26	100	20	100	78	100

Chi Squared = 2.86 P = 0.239

Other forms of technology acquisition

There are other means of acquiring technology through more formal forms of technology transfer. In particular, both licensing and joint ventures, in principle, offer relatively economical means of product acquisition when compared with independently produced products from within internal R&D departments. However, in keeping with other research into high-technology small firms performed by this author (Oakey et al 1988; Oakey et al 1990), only 23 (17%) of all survey firms maintained licensing agreements with other firms, while a smaller group of 21 firms (16%) had established joint ventures with other enterprises. Neither in the case of licences or joint ventures was there any particular sectoral bias in these minority results. Licences are often less attractive to high-technology small firms in practice, since a licensed product is often not "state of the art" due to its original exploitation by the licensing firm. Moreover, joint ventures, while attractive in principle due to their "risk sharing" nature, are often difficult to control organisationally, in terms of agreeing both equitable inputs of investment capital, and a fair sharing of the intellectual property outputs produced.

Employment problems and internal innovation

It is inevitable in high-technology small firms, where a high proportion of the staff are directly or indirectly engaged in the development and manufacture of technically sophisticated products, that most problems associated with labour shortages will involve technical staff. Clearly, if key workers are not obtainable from the local or national job market, training will be the only alternative for firms wishing to increase their R&D capacity in competitive conditions. The above data on numbers of R&D workers employed bear witness to the substantial numbers of technically qualified workers survey firms may require. The evidence presented below examines if there was a problem of labour shortages, whether any observed shortages inhibited the innovative performance of survey firms, and the steps (if any) firms had taken internally to train skilled staff.

TABLE 4.15 *Type of worker shortage by sector*

Type of worker	Biotechnology		Electronics		Software		Total	
	N	%	N	%	N	%	N	%
Scientist and R&D	16	76.2	10	47.6	10	45.5	36	56.3
Skilled shopfloor	2	9.5	4	19.0	4	18.2	10	15.6
Management	1	4.8	3	14.3			4	6.3
Other	2	9.5	4	19.0	8	36.4	14	21.9
Total	21	100	21	100	22	100	64	100

Chi Squared = 10.24 P = 0.115

The issue of skill shortages was initially approached by asking respondents in the three study sectors if they had experienced difficulties in recruiting any particular type of labour in the past year. It was to some extent surprising to discover that the substantial minority of 47% of survey firms noting labour recruitment difficulties were not biased in favour of any particular geographical region or industrial sector. However, more detailed investigation of the type of worker causing the problem of labour shortage in Table 4.15 conclusively indicates that the category "scientist/ R&D worker" was the major problem category in 56% of cases. As might be expected, this problem was particularly acute in the biotechnology firms, where this category of labour shortage represented 76% of responses.

To assess the impact of these shortages on the R&D performance of survey firms, respondents were asked if the noted shortages had directly inhibited innovation. Table 4.16 indicates that 53% of responding firms (a minority of the survey total at 20%) claimed that labour shortages had directly inhibited their internal innovation performance. The extent of this minority response is somewhat surprising, given that it emanated from a survey taking place during a severe recession. Moreover, Table 4.16 further indicates, as might have been anticipated, that the problem of labour shortages inhibiting innovation was greatest among the firms with the highest growth in turnover. A large majority of 68% of firms experiencing over 100% turnover growth and acknowledging labour shortages claimed that they inhibited their innovation performance. Although it might appear to be a contradiction that firms with high turnover growth should claim *inhibition* of their innovative performance by labour shortages, it is often the fastest growing firms that put most stress on their local and national labour markets, while it is clearly possible that the high growth rates recorded in such firms might have been yet greater, had adequate labour been available.

TABLE 4.16 *Recruitment problems an inhibition to innovation by growth in turnover (1987-91)*

	Under 25%		25-100%		Over 100%		Total	
	N	%	N	%	N	%	N	%
Inhibition to innovation	4	44.4	5	31.3	19	67.9	28	52.8
None	5	55.6	11	68.8	9	32.1	25	47.2
Total	9	100	16	100	28	100	53	100

Chi Squared = 5.78 P = 0.055

Given that a substantial number of firms had noted labour shortages, it appeared appropriate to ask if internal training programmes were operated by survey firms to solve skill shortages. Table 4.17 reflects a generally impressive level of training provision in survey firms, with 76% of all establishments providing "in-house" training. Interestingly, an almost statistically significant difference is produced in Table 4.17 by a sharp difference between an almost identical performance for biotechnology and electronics respectively at 82% and 83% levels of internal training provision, and software, where the level of "in-house" training was a lower 64%. An explanation for this lower level of provision in the software sector might be that programming skills are the major form of software expertise, and the high level of graduate employment in this sector tends to mean that many new employees arrive at survey software firms with a high level of appropriate training. The need for adapting basic skills to specific firm needs are more acute in the electronics and biotechnology sectors, where technological specialisms within firms are far more diverse.

TABLE 4.17 *Incidence of internal training by sector*

	Biotechnology		Electronics		Software		Total	
	N	%	N	%	N	%	N	%
Training	36	81.8	38	82.6	30	63.8	104	75.9
None	8	18.2	8	17.4	17	36.2	33	24.1
Total	44	100	46	100	47	100	137	100

Chi Squared = 5.72 P = 0.057

TABLE 4.18 *Incidence of internal training by percentage employment growth (1987-91)*

	Under 25%		25-100%		Over 100%		Total	
	N	%	N	%	N	%	N	%
Internal training	27	62.8	30	71.4	39	92.9	96	75.6
None	16	37.2	12	28.6	3	7.1	31	24.4
Total	43	100	42	100	42	100	127	100

Chi Squared = 10.99 P = 0.004

Table 4.18 tends to support earlier evidence by strongly indicating that it is the fastest growing firms that have the most positive approach to the labour market, here reflected in their attitude towards internal training. In particular, 93% of firms in the over 100% employment growth category maintained internal training provision, compared with a 63% level in the less than 25% growth category, a difference that was significant at the p = 0.004% level. However, it should be stressed that a 93% level of internal training is exceptional, and that the average 76% level of internal training is very impressive. This high level of internal training activity in the midst of a recession, when money is "tight" and training is often an early casualty of hard times, is particularly heartening. It is almost certainly a reflection of the essential nature of skilled workers, *appropriately trained* to the specific needs of NTBFs. In firms where "niche" production is prevalent, workers who have been trained for the specific tasks required by such firms are preferable to new employees with generally relevant skills, that subsequently must be adapted (Oakey 1981). Thus, a high level of internal training must be partly a reflection of the specialised needs of survey firms, especially in the biotechnology and electronics sectors.

Summary conclusions

The introductory assertion of this chapter that high-technology small firms would be generally dependent on a high level of "front-end" R&D funding to allow the development of sophisticated products and services has been largely substantiated. The overall high incidence of an "in-house" R&D capability, in the form of a full-time staff of R&D workers and consequent R&D expenditure, was direct confirmation of the required capital investment in R&D, which was the cause of many of the funding needs detected in Chapter 3 above. However, although the generally high level of R&D investment in survey firms had been expected, the financial extent and temporal intensity of R&D investment in a significant minority of biotechnology firms was at an *extremely high* level that was surprising.

Thus, differences in R&D investment between NTBFs are subtle in the sense that they are not manifested in terms of a normal range of commitment from high to low investment. Instead, in the case of these survey firms, there are sharp differences, but they appear to be *within* the generally high end of an R&D investment continuum. None the less, from the viewpoint of identifying high-technology small firms and designing optimal strategies for their promotion, it would be wrong to assume that, because most high-technology small firms invest heavily in R&D, they can be treated as a common group. The extremely high levels and intensities of R&D investment in many biotechnology firms are of an intensity that renders them special cases, even among a generally high-technology group of firms. External investors contemplating involvement in such firms must employ a different set of criteria when contemplating amounts of required investment, pay-back periods, risks and, as mentioned in the introduction above, the mitigating potential for profitability *if* the technology developed by the firm is an ultimate success.

A further recurrent theme of this book, supported by this chapter, has been the propensity for there to be a sectorally-based pattern of results during analysis. Certainly in this examination of R&D commitment, there was a frequent trend that indicated a high level of activity for biotechnology firms, with a moderate performance among electronics firms, concluding with the weakest level of R&D activity in the software enterprises. This underlying pattern of differing formation and early growth costs among these generally high-technology small firms continues to be supported by this evidence on R&D commitment.

Although it might be argued that such diverse patterns of investment and performance are due to variations in management competencies, it is far more likely that, in accordance with the general thesis of this book, technological constraints overwhelm any varying impact caused by differences in managerial quality to produce a pervasive deterministic impact on the manner in which firms may be established and developed.

CHAPTER 5

Purchasing, Sales and Marketing

Introduction

This chapter contains a consideration of three important functional aspects of high-technology small firm operation. Much of the preceding discussion has been justifiably concerned with the internal operation and growth of newly formed firms. However, this investigation of material inputs, outputs and marketing is designed to emphasise the important role of external materials or networks in determining the successful functioning of a new high-technology enterprise. The efficient procurement of materials with which products are manufactured, and the dispatching of the final product to customers, is a key determinant of success in any industrial firm. Indeed, proximity to a particularly rich and abundant set of actual and potential suppliers has long been acknowledged to be a major component of agglomeration economies in traditional industries (Marshall 1890; Hall 1962; Martin 1966; Wood 1969), offering locational advantage in particular regional contexts (e.g. the cotton textile areas of North West England and New England in the United States).

For an extended period after the Second World War, there was a pervasive belief that the modern industry of the 1950s and 1960s had become generally "footloose", and that older concerns for the constraining influence of local material inputs to the production process were not relevant to the relocating industries that were the currency of mobile industry policy, designed to achieve relocations away from the (then) booming South East and West Midlands of England. However, the claim that all modern industries were "footloose" and consequently insensitive to the economies afforded by a rich clustering of local suppliers, was an over-generalisation. While the relocation of some forms of production was achieved without detriment (e.g. electrical "white goods" manufacture) (Luttrell 1962), relocation of motor vehicle production away from its supplier base in the West Midlands of England was less successful (e.g. the demise of the car assembly plant at Linwood outside Glasgow). Moreover, studies of recently relocated firms tended to suggest that the "friction" of distance was a significant factor in reducing the efficiency that firms enjoyed with their suppliers, when compared to their previous location (Townroe 1971). Such diseconomies were associated not merely with a loss of physical efficiency due to the increasing geographical distance caused by the move, but was also linked with increased *organisational* friction due to an inability of relocated firms to achieve regular "face to face" contact with their traditional suppliers. The need for such frequent personal interaction was, not surprisingly, associated with increased complexity of production (Wood 1969).

The emergence of the high-technology industrial growth poles in the 1970s (e.g. Silicon Valley; Route 128 in the United States) further supported the view that many of the most modern firms were not "footloose", but were conversely heavily locationally dependent on skilled labour, sophisticated localised material inputs and local financial support (Cooper 1970; Oakey 1984). For many sophisticated high-technology small firms, a Silicon Valley location represented a significant advantage, both in terms of competitive prices obtained from a large number of competing local suppliers, and through the highly sophisticated technology incorporated in the goods and services provided by such local specialist suppliers. However, not all the attributes of an agglomerated location were positive. The popularity of such locations caused congestion that resulted in the negative phenomena associated with "overheating", such as high site and labour costs.

The main lesson to be learnt from all these experiences is that it is always dangerous to generalise about the locational requirements of manufacturing industry in general, or high-technology industries in particular. Indeed, regarding high-technology small firms, it would be wrong to assume that, merely because such firms gain agglomeration economies from local supplier firms, their material inputs are consequently mainly local in origin. While the sophisticated needs of high-technology small firms ensure that the existence of specialist local suppliers is an advantage, it is also the case that only rarely can such firms source *all* their, often complex, material inputs from the local area. Indeed, it is of significance to this current consideration of high-technology new firm formation and growth to note that many of the inputs to these firms are international in origin.

A number of key features of high-technology small firm production ensure that the input (purchase) and output (sales) linkages tend to be organisationally and geographically complex in nature. Not only are inputs to the production process geographically complex due to a general need for many complex input materials and sub-assemblies, but also the complex and specialist nature of final products engender customers who are highly specialised and geographically diverse. For example, a high-technology small firm manufacturing a sophisticated pressure measuring instrument for the aircraft industry would source many of its material inputs from local *and* international suppliers, while its market typically also would be internationally located. This feature is particularly prevalent in high-technology industries for two major reasons. First, quite simply, manufacturers of low-technology products have no need to incur excessive procurement costs by purchasing from exotic and comparatively expensive international customers. Especially for bulky low value input materials, it is inevitable that the most competitive price will be obtained locally from easily available suppliers of ubiquitous inputs. Second, a more complex explanation for the international procurement of material inputs among high-technology small firms is that, although it is well established that such firms *need* to source and sell internationally, they are only able to do so as a result of the *high profit margins* achievable during manufacture. These high margins are critical in that they permit both the relatively expensive procurement of inputs to the production process and distribution to the customer, while ensuring that there is an attractive residual net profit after such high expenses have been incurred. The main reason for the achievement of such high profit margins is the high value added in production provided by exclusive intellectual property of the high-technology small firm concerned. Often the ability of such firms to solve specific problems in niche areas of production ensures that very high prices can be charged. This situation contrasts strongly with other areas of ubiquitously known production (e.g. printed circuit board making), where many entrants to a subsequently highly competitive market ensure that profit margins are narrow.

Of particular interest to this book on the formation and growth characteristics of high-technology small firms in differing industrial sectors is a new phenomenon that has recently emerged in high-technology production. There has been a gradual evolution throughout the 1980s in which NTBFs have grown from their original instruments-electronics-semiconductor industry base, to include newer forms of production where small innovative firms are important. Significantly, high-technology small firms in this study derived from the biotechnology and software sectors are appropriate examples of this new type of firm that occupies the interface between manufacturing and service industries. A key characteristic of such sectors is that a larger proportion of the value added in their products is contained in intellectual value added than is the case for electronics-based high-technology small firms. When compared with other low-technology forms of production (e.g. mechanical engineering), electronics-based high-technology small firms manifest a high level of intellectual value added in their products. However, such new small firms from the emerging biotechnology and software sectors rely *even less* on material and sub-assembly inputs for their value added. In the biotechnology industry, a new biomedical product is often represented by a vial of liquid in which the material value is virtually nil, but the intellectual value added may be massive. Similarly, in the case of software, the addition of a software programme to a disk worth less than one pound, can increase its value to make it worth thousands or tens of thousands of pounds, with no gaining of weight. In such circumstances, the contribution of intellectual value added to the total value added of the product (including materials and all other costs) is virtually one hundred per cent.

Recent studies have noted that this increasing proportion of intellectual value added to the total value added of products has permitted a new locational freedom to firms enjoying such benefits, if they *wish* to take advantage of such potential freedom (Oakey and Cooper 1989). These newly emerging firms, with major intellectual inputs to the production process and few material requirements, are frequently freed from dependence on material suppliers. This renders such firms truly "footloose", and a trend has emerged in which a significant minority of NTBFs in recent studies were located in peripheral rural locations (Oakey et al 1990; Keeble 1994). There is a well established body of literature to testify to the tendency among firm founders to build "psychic income" into consideration of where to locate (Greenhut 1956; Cyert and March 1963), a practice in which such "psychic income" is substituted for financial income when an entrepreneur chooses an environmentally attractive plant location in preference to a more economical production location. Put simply, these arguments suggest that, for firms with a high level of intellectual value added, there is little *economic* difference between agglomerated urban or peripheral rural locations, rendering the choice of an attractive rural location a small price to pay in efficiency terms, while maximising psychic income.

All the above comments on input and output linkages (or networks) suggest that, not only do linkages vary in importance *between* high- and low-technology sectors, but they may vary considerably in their nature and impact *within* high-technology sectors. Again, this study benefits from a data set that is ideally equipped to explore the varying impact of input and output linkages on new high-technology small firms in the three study sectors. The above observations provide a number of interesting issues to be explored in the empirically-based analytical sections below. Initially, analysis will focus on the impact of supply linkages on firm formation and growth, followed, in keeping with the flow of materials through a production plant, by a consideration of the impact of customers and marketing to customers on the operation and performance of survey firms.

Purchasing patterns

Input materials

It is likely that, given the opportunity, most firms would choose to purchase material inputs to the production process from local sources. Given satisfactory quality and price, the proximity of a local supplier renders frequent contact through telephone conversation or personal visits more logistically viable at reduced cost, when compared with more distant potential suppliers. Frequent interaction with suppliers is generally more prevalent among high-technology small firms, due to the often complex nature of both the materials supplied (which are usually products in their own right [e.g. transformers; circuit boards]) and the products into which they are included. The great strength of Silicon Valley, and in a less concentrated manner, the South East of England, is that a great variety and choice of such specialist suppliers exist in a discrete geographical area. Indeed, it is often the absence of such a rich supply network that is a major barrier to high-technology growth in a declining region, historically dependent on traditional low-technology industries (e.g. textiles; coal mining; heavy engineering) (Oakey 1984).

In the British context, the comments on the possible concentration of high-technology supply linkages in South East England is supported by Table 5.1, which indicates the percentage of material inputs (by value) derived from within a 30-mile radius of the plant. Apart from the initial overall observation that a surprisingly small 37% minority of survey firms derive over 25% of their inputs locally, it is significant to note that the regional distinction between the South East and East Anglia and the rest of Britain clearly indicates that, while 49% of the South East and East Anglian firms derived over 25% of their input from within 30 miles, the equivalent figure for the rest of Britain was 25%, significant at the $p = 0.007\%$ level during Chi Squared testing (Table 5.1). This result, provided by an amalgam of three high-technology sectors, suggests that the South East may be a beneficial location from the point of view of material supply to high-technology industrial firms in keeping with earlier work (Oakey 1984; Oakey and Cooper 1989).

TABLE 5.1 *Percentage of materials derived from within 30 miles by regional location*

% Materials within 30 miles	Rest of Britain		East Anglia/ South East		Total	
	N	%	N	%	N	%
< 25%	47	74.6	29	50.9	76	63.3
> 25%	16	25.4	28	49.1	44	36.7
Total	63	100	57	100	120	100

Chi Squared = 7.254 P = 0.007

Interesting results are produced from an analysis of the percentage of material inputs purchased locally when disaggregated to the level of the three survey sectors in Table 5.2. While biotechnology and electronics firms show a strong propensity to purchase outside the local area, with only 23% and 33% respectively of these firms purchasing over 25% of their inputs locally, the figure for software was much higher at 59%, showing significant statistical difference through Chi Squared testing at the p = 0.005% level. This discrepancy between sectors is probably explained by the low value of total material inputs to the production process within the software industry. Since much of the value added in production relates to intellectual value added (argued above), the contribution to final product sales value of material inputs is thus relatively low when compared with the electronics and biotechnology sectors. Of the relatively small amount of materials purchased from outside sources, much can be purchased locally (e.g. stationery; disks) although, as noted below, a small proportion of inputs are purchased from abroad (e.g. specialist programmes and hardware etc.). The general trend of these results confirms earlier work on high-technology small firms (Oakey 1984), which conveys a mixed picture where firms may ideally seek to source inputs from the local area and gain benefits from such local sourcing when possible. However, the complexity of production ensures that, even in locations of high-technology agglomeration (e.g. Silicon Valley), a substantial proportion of material inputs must be derived from distant (often foreign) locations.

TABLE 5.2 *Percentage of materials derived from within 30 miles by sector*

% Materials within 30 miles	Biotechnology		Electronics		Software		Total	
	N	%	N	%	N	%	N	%
< 25%	33	76.7	30	66.7	13	40.6	76	63.3
> 25%	10	23.3	15	33.3	19	59.4	44	36.7
Total	43	100	45	100	32	100	120	100

Chi Squared = 10.6 P = 0.0048

TABLE 5.3 *Percentage of materials directly imported by sector*

% Materials imported	Biotechnology		Electronics		Software		Total	
	N	%	N	%	N	%	N	%
0	7	17.1	18	39.1	22	73.3	42	40.2
1-25	14	34.1	18	39.1	2	6.7	34	29.1
>25%	20	48.8	10	21.7	6	20.0	36	30.8
Total	41	100	46	100	30	100	117	100

Chi Squared = 27.5 P = 0.00002

Indeed, the length of supplier linkages is further explored in Table 5.3 which seeks to discover the proportion of material inputs (by value) derived from abroad through imports. A very clear distinction may again be drawn between sectors, with a transition from biotechnology, where 49% of firms derived over 25% of their inputs by value from abroad, which fell to 22% for electronics firms and 20% for software companies. These sectoral differences were again highly significant through Chi Squared testing at the p = 0.00002% level. Probably the most realistic means of interpreting these sectoral differences is to observe that the percentage of imports for biotechnology study firms is *particularly* high, while the other two sectors would also be high if judged against other low-technology sectors (Oakey et al 1990). These results confirm the observation that it is often the case that high-technology small firms cannot obtain a significant proportion of the sophisticated raw material and sub-assemblies they require for their production from the local area, and are forced to source material from distant national locations, or from abroad.

TABLE 5.4 *Main country of origin by sector*

	Biotechnology		Electronics		Software		Total	
	N	%	N	%	N	%	N	%
EC	2	6.3	4	14.3			6	8.8
Europe	4	12.5	2	7.1			6	8.8
US/Canada	18	56.3	10	35.7	6	75.0	34	50.0
Far East			5	17.9	2	25.0	7	10.3
Many places	8	25.0	7	25.0			15	22.1
Total	32	100	28	100	8	100	68	100

Chi Squared = 14.1 P = 0.08

Table 5.4 indicates the countries cited by survey respondents as the main sources of imports to their firms. It is perhaps surprising, given the generally strong export orientation of British firms to Europe, that neither the EC nor the rest of Europe featured as major import locations (i.e. both receiving 9% of responses). Conversely, North America was a major source location for all sectors (i.e. 50% of all respondents), but was especially popular with biotechnology and software respondents with respective 56% and 75% levels of acknowledgement. The strong link held by small British biotechnology firms with North America has been noted elsewhere (Oakey et al 1990), and largely stems from the strength of the emerging biotechnology industries in these locations. The strong software orientation to the United States in the minority of cases registered in Table 5.4, is largely related to specialist hardware and software purchase of equipment to facilitate the software production process.

Sub-contracting

Sub-contracting is a specialised form of input to the production process in which manufacturing firms "put out" work to sub-contractors which would otherwise need to be performed "in-house". Such behaviour offers a number of advantages to the sub-contractor, to weigh against the loss of value added involved in the surrender of a part of the manufacturing process. First, the sub-contractor may be able, due to specialist equipment, to perform the work more economically and to a higher standard than could be achieved by the sub-contracting firm. Second, as noted in the case of sub-contractors to high-technology small firms in Silicon Valley, the sub-contractor may provide technical expertise that is lacking in the sub-contracting firm that enables the removal of a technical bottleneck in the design and/or production of a component (Oakey 1984). Third, sub-contracting is a very flexible strategy in that work can be expanded or contracted by the use of sub-contractors without the need to hire or fire workers or alter internal working methods of the commissioning firm.

However, the extent to which the attributes of the sub-contracting process are attractive to the diverse high-technology sectors represented in this study will be of particular interest. Initially, it might be argued that the dependence on sub-contractors in general, and local sub-contractors in particular, will largely depend on the degree to which the value added in production in individual sectors is determined by intellectual or material value added. As intellectual value added increases, it might be expected that the importance of material inputs to the production process through sub-contracting would decline. Thus, local sub-contracting might be anticipated to be higher in electronics firms than in their software counterparts. Although it is possible for firms producing a high level of intellectual value added to sub-contract intellectual services such as R&D, it is generally easier to sub-contract out the manufacture of hardware (which can be modularised and self-contained) than the development of ideas.

Such conjecture is supported by data in Table 5.5 where there is clear evidence, statistically significant at the $p = 0.009\%$ level during Chi Squared testing, that the electronics firms of the study *are* more likely to sub-contract out parts of their production. While 72% of these electronics firms sub-contracted proportions of their manufacturing, the figure for biotechnology and software firms was 59% and 40% respectively. This pattern of results broadly conforms to the amount of material value added in production for these three sectors, with electronics highest and software the lowest. Table 5.6, which gives some indication of the proportion of the local sub-contracting "put out" by firms within a 30-mile radius of the factory, again is statistically significant at the p = 0.036% level, and indicates that only 19% of electronics firms sub-contracted less than 25% of their work locally, compared with a much higher 54% and 39% for biotechnology and software firms respectively. These percentage results should be interpreted with Table 5.5 in mind, since it should be remembered that less than half of software firms (i.e. 40%) sub-contracted at all. However, although sub-contracting in general and local sub-contracting in particular was highest among electronics firms, a minority of firms in all sectors put out all their work within the local area (Table 5.6), thus confirming the popularity of local sub-contracting arrangements, when possible.

TABLE 5.5 *Incidence of production/services sub-contracted out by sector*

	Biotechnology		Electronics		Software		Total	
	N	%	N	%	N	%	N	%
Sub-contracted out	26	59.1	33	71.7	19	40.4	78	56.9
None	18	40.9	13	28.3	28	59.6	59	43.1
Total	44	100	46	100	47	100	137	100

Chi Squared = 9.42 P = 0.009

TABLE 5.6 *Percentage sub-contracted out within 30 miles by sector*

% sub-contracted within 30 miles	Biotechnology		Electronics		Software		Total	
	N	%	N	%	N	%	N	%
<25%	14	53.8	6	18.7	7	38.9	27	35.5
25%-99%	3	11.5	14	43.8	5	27.8	22	28.9
100%	9	34.6	12	37.5	6	33.3	27	35.5
Total	26	100	32	100	18	100	76	100

Chi Squared = 10.27 P = 0.036

Survey firms were also asked if any of their sub-contractors were located internationally since earlier work had indicated that biotechnology firms had maintained international sub-contractors (Oakey et al 1990). A relatively small minority of survey firms acknowledged international sub-contractors. Six software firms, three biotechnology firms and two electronics firms (eleven in total) claimed this form of supplier. However, perhaps of more importance than sector was the discovery that a majority eight of these firms recorded unit profit margins of over 50%. Evidence from this minority of survey firms tends to support the above introductory assertions of this chapter that firms with high profit margins are *allowed* relatively exotic supply chains due to the high value added of the products they produce. It is also probably worth restating that, in a number of cases, such inputs are essential to the successful production of products. Indeed, exotic subcontractors would be unlikely to gain preference over local suppliers if the required service type and/or quality was available locally.

Customer patterns

The existence of important local customers is a second key component of agglomeration economies discussed in the introduction to this chapter. In all traditional agglomerations, the total agglomerative effect was comprised of interaction between firms as vertically disintegrated production was shared between independent producers. Goods and services passed along local chains of production as value was added by each independent specialist. Typical traditional industries in sub-regional economies often displayed this agglomerative phenomenon which has been noted for a number of industries and geographical regions (e.g. guns and jewellery [Wise 1949]; furniture [Hall 1962]; scientific instruments [Martin 1966]). Consequently, the markets of the small supplier satellite firms were almost exclusively local. Recent work on Silicon Valley in California has noted that this general phenomenon has re-emerged in high-technology agglomerations, with one new feature (Oakey 1984). Although many of the small Silicon Valley electronics firms sell a significant proportion of their output locally within the agglomeration (thus creating local agglomeration economies), unlike the traditional industries discussed above, they also export a varying, but substantial, amount of their output to locations *outside* the region.

Unlike industries of the Victorian era, modern high-technology agglomerations are part of an international network of production in which the high value added in production of much of their outputs permits sales to international customers. Previous research has indicated that a combination of insignificant transport costs as a proportion of high unit sale prices and the international spread of specialist customers renders exporting a necessity for most high-technology small firms in general, and new biotechnology firms in particular (Oakey 1984; Oakey et al 1990).

However, another *strategic* reason for geographical, sectoral and individual diversity of customers (understood but not available to most Victorian producers) is the desire to obtain the safety inherent in "not putting all your eggs in one basket". While relatively parochial Victorian producers in agglomerations were often unable to obtain customers outside the single industry dominated agglomeration, modern high-technology small firms, with their strong export potential, are often reluctant to send a large proportion of their output to a single large customer. The sudden closure of this patron, a decision to internalise work previously "put out" to the small firm concerned, or loss of such an important market to a competitor could, in extreme circumstances, cause the demise of the small supplier firm concerned. Conversely, a wide geographical and sectoral spread of customers is a good hedge against any particular regional or sectoral slump in activity. Indeed, the great strength of all agglomerations also holds their potential downfall, since the very specialisations that create comparative advantage today, render them vulnerable if such a regional specialism comes under threat from an external source. The decline of the cotton textile industry of North West England is a classic example of this phenomenon.

Incidence of a single major customer industry

None the less, although high-technology small firm managements may desire a diverse customer base, difficult trading circumstances in times of recession may render such prudence an unachievable goal. When orders are scarce, any patron will be acceptable, regardless of their share of total sales. This observation may be relevant, since the survey work discussed below took place during a period of growing recession in Britain. Initial evidence on the sectoral dominance of the sales of survey firms is provided in Table 5.7, where it is clear that all three survey sectors displayed a strong tendency for a major customer sector purchasing over 25% of their output by value, with an average across the sectors of 75%. However, the percentage for biotechnology firms was an *extremely* high 89%, which caused significant difference during Chi Squared testing of p = 0.032%. A further investigation of the industries to which this high share of output was dispatched did not reveal any particular clustering by sector.

TABLE 5.7 *Incidence of a single industry buying over 25% sales by sector*

	Biotechnology		Electronics		Software		Total	
	N	%	N	%	N	%	N	%
Single industry	39	88.6	31	67.4	32	68.1	102	74.5
None	5	11.4	15	32.6	15	31.9	35	25.5
Total	44	100	46	100	47	100	137	100

Chi Squared = 6.86 P = 0.032

This dispersed result was particularly surprising in the biotechnology instance, since it had been expected that the chemical and pharmaceuticals industries would be particularly important customers for the new small biotechnology firms. However, a regional analysis of the location of these customer industries purchasing over 25% of survey firms' output revealed strongly divergent sectoral results. The international nature of markets in the biotechnology survey firms is strongly confirmed in Table 5.8 in which 70% of the dominant industries acknowledged by the biotechnology firms were located abroad, compared with 26% for electronics firms and 28% for software producers. Such sectoral difference was sufficient to produce a strongly significant statistical difference through Chi Squared testing at the p = 0.0005% level. This clear evidence on sales, taken together with the above evidence on input materials, confirms the wide geographical spread of input and output linkages in the small biotechnology firms of this study.

TABLE 5.8 *Location of industry with over 25% of sales by sector*

	Biotechnology		Electronics		Software		Total	
	N	%	N	%	N	%	N	%
Same region	1	2.5	6	19.4	7	21.9	14	13.6
National	11	27.5	17	54.8	16	50.0	44	42.7
Abroad	28	70.0	8	25.8	9	28.1	45	43.7
Total	40	100	31	100	32	100	103	100

Chi Squared = 20.01 P = 0.0005

TABLE 5.9 *Incidence of a customer at formation buying over 10% by sector*

	Biotechnology		Electronics		Software		Total	
	N	%	N	%	N	%	N	%
Customer purchasing >10%	21	50.0	34	77.3	29	61.7	84	63.2
None	21	50.0	10	22.7	18	38.3	49	36.8
Total	42	100	44	100	47	100	133	100

Chi Squared = 6.93 P = 0.0312

Incidence of a single major customer firm

The problem of over-concentration on a single industry that might suddenly encounter difficulties is further accentuated if such a large proportion of sales is dispatched to a single customer that might experience severe difficulties. Thus survey firms were asked to indicate both at the time of formation, and in 1991, if they possessed a single customer purchasing over 10% of their output (by value). Table 5.9 indicates that some sectoral difference existed in terms of the first year after formation when respondents were asked whether there was a single major customer. While electronics and software firms respectively recorded 77% and 62% levels of single customer acknowledgement, the proportion for biotechnology firms was a lower 50%, causing significant difference during Chi Squared testing at the $p = 0.031\%$ level. The high average level of 63% acknowledgement probably reflects the reality that firms are often triggered into existence by the potential patronage of a single major customer, which may be the large "incubator" firm from which the new firm "spun off". However, the position for more established firms in 1991 changed only marginally in that while the sectoral average for a single major customer reduced only slightly to 58%, the biotechnology firms fell into line with the other two sectors by experiencing an increase in the proportion of firms with a single major customer from the 50% recorded above to 59%. This relatively high level of dependence on a single major customer may, to some extent, be explained by the reality, suggested above, that a broadly spread customer base is only a goal. In the recent recessionary conditions, such ideal of diversity may not be a realistic objective for many firms, and survival on the basis of *any* sales may be a higher priority.

While the proportion of firms claiming a single major customer purchasing over 10% of sales (by value) showed little change between formation and 1991, there was an overall decline in the percentage of sales to such major patrons over the period. Although not warranting tabulated presentation, a surprisingly high half of all survey firms in the first year of formation (with little sectoral divergence) dispatched over 50% of their output to a single major customer. By 1991, the proportion of firms in this over 50% sales to a single customer category had declined to 22%. However, these results on the *extent* of single customer sales, taken together with the results on the extent of important single customers, tend to suggest that many of the new firms in all sectors are rather precariously dependent on single customers, implying all the dangers discussed above.

TABLE 5.10 *Location of customer purchasing over 10% of output by sector in 1991*

	Biotechnology		Electronics		Software		Total	
	N	%	N	%	N	%	N	%
Same region			8	30.8	12	42.9	20	25.3
National	2	8.0	14	53.8	11	39.3	27	34.2
Abroad	23	92.0	4	15.4	5	17.9	32	40.5
Total	25	100	26	100	28	100	79	100

Chi Squared = 42.17 P = 0.00001

Table 5.10 on the location of a single major customer is partly influenced by the results given in Table 5.8 on the location of industries purchasing over 25% of a firm's output, since a single major customer *may* explain the importance of a single industry to the survey firm concerned. It is certainly clear that the high level of international linkages recorded by biotechnology firms in Table 5.8 is continued in Table 5.10. Biotechnology firms with a single major customer abroad rose from a comparatively high 57% in the year of formation to a striking 92% in 1991 (Table 5.10), while the other two sectors only recorded marginal increases over the same period. Throughout, the greater international orientation of biotechnology firms was maintained, such that in 1991 the 92% level of firms with an international major customer can be compared to 15% for electronics firms and 18% for software enterprises. This difference was strongly significant during Chi Squared testing at the p = 0.00001% level. Such a geographically dispersed set of customer relationships suggests that the high value added of the biotechnology industry, and the complex technologies inherent in the products of the firms concerned, respectively both allow and provide a reason for such long distance international trade. While the economic viability of long distance export sales is *necessary* for the achievement of international sales, the demand from equally specialist international customers provides the *sufficiency*. Conversely, it might also be noted that the lower level of international sales among electronics and software firms is partly explained by their ability to find customers in their local regional or national environment, thus *negating the need* for complex international sales relationships.

Exports

As might be expected from the above evidence on customer locations, survey firms displayed a high level of exports, with all firms exporting to some extent, and 79% of these enterprises exporting more than 25% of their output. From a sectoral viewpoint, exports (by value) were clearly higher in the biotechnology sub-group of firms. While 62% of biotechnology firms exported 50% or more of their output, the equivalent figure for electronics and software firms was 36% and 24% respectively (Table 5.11). This sectoral difference again produced strong significance during Chi Squared testing at the p = 0.0001% level. The extreme export orientation of the biotechnology firms in this study, while clearly worthy of note, is in danger of creating an impression that the two remaining sectors are poor exporters. However, the true picture created by all the customer information in this sub-section is that all three industrial sectors are generally highly export oriented. The performance of the biotechnology survey firms is merely an *extremely* high export performance within an overall high level of export orientation.

TABLE 5.11 *Percentage of output exported by sector in 1991*

%Exported	Biotechnology		Electronics		Software		Total	
	N	%	N	%	N	%	N	%
10-24%	3	7.1	6	13.6	19	41.3	28	21.2
25%-49%	13	31.0	22	50.0	16	34.8	51	38.6
50%+>	26	61.9	16	36.4	11	23.9	53	40.2
Total	42	100	44	100	46	100	132	100

Chi Squared = 24.3 P = 0.0001

Sales and marketing

Although, in principle, sales and marketing are key final functions in the total innovation process that greatly influence the profitability of a product innovation, previous work has clearly indicated that many high-technology firms neglect marketing and merely rely on unsolicited orders (Oakey et al 1988). The relationship of high-technology small firms to their customers is strongly influenced by the often complex nature of the purchased product or service. Unlike a consumer durable, where the technology and its function is established, and where successful operation can be achieved with a relatively simple service manual, much of the value added in high-technology small firm products relates to the technical service that accompanies the product in circumstances where the customer is often another manufacturing firm, and not a final consumer. Thus, the customer firm often buys the technological expertise of the NTBF in order to provide a technological competence that it cannot or will not produce internally. Indeed, the availability of these input materials and sub-assemblies is a major agglomerative advantage of Silicon Valley (Oakey 1984).

In this context, sales staff in high-technology small firms not only perform the normal selling function, but often liase with existing customers to provide after sales service on past product sales. A further function performed by sales staff is the gathering of "intelligence" on shortcomings in previous products to customers, and suggestions for future modification to existing products, and for totally new products, that might be of more general relevance to the whole market. None the less, such theoretical advantages, while of clear value in the long term, may be forgone by NTBFs because of short-term financial problems. Given adequate demand from unsolicited sales, many high-technology firms decide to avoid the risk associated with expensive marketing costs, where the return on such investment cannot be predicted, and virtually allow the product to "sell itself" through reputation, with only a minimal advertisement in selected trade journals (Oakey et al 1988). The comparison of sales and marketing activity across the three study high-technology sectors below should provide interesting insights into the above observations.

Main product promotion methods

As an initial step in this consideration of marketing and sales, survey firms were asked to state the primary method by which sales were achieved. The generally low emphasis placed on marketing in a substantial minority of survey firms in all sectors is reflected in the 27% of firms noting "word of mouth" as the main unsolicited method by which goods and services were sold (Table 5.12). Normally, this form of selling involves the reputation a product achieves as it is distributed among customers in a narrow "niche" market. A prime example of this phenomenon is the spread of new medical instruments through a "grapevine" of consultants within hospitals.

TABLE 5.12 *Main sales promotion method by sector*

	Biotechnology		Electronics		Software		Total	
	N	%	N	%	N	%	N	%
Word of mouth	11	27.5	9	20.5	15	31.9	35	26.7
Adverts	3	7.5	10	22.7	8	17.0	21	16.0
Mail shots	4	10.0	3	6.8	6	12.8	13	9.9
Conference/ exhibitions	6	15.0	6	13.6	9	19.1	21	16.0
Visits	13	32.5	13	29.5	5	10.6	31	23.7
Other	3	7.5	3	6.8	4	8.5	10	7.6
Total	40	100	44	100	47	100	131	100

Chi Squared 10.90 P = 0.365

Marginally more proactive is the use of advertisements in 16% of survey firms, mainly in selected "niche" trade journals, where specialist consumers are likely to seek these often specialist products and services. Conferences and exhibitions received equal support at 16%, which is a more personal means of reaching specialist customers in a forum where new products may be explained and demonstrated. The final specific means of reaching customers is by means of formal sales visits from firm staff. As will be discovered below, the use of sales visits by dedicated staff is associated with high growth firms, although the relationship between these variables is difficult to establish.

Commitment of resources to sales and marketing
The information given in Table 5.12 above suggests that there may be wide variations in the amount of resources that different survey firms devote to sales and marketing, ranging from very little for those firms relying on "word of mouth", to substantial sums where sales visits are concerned. This strong divergence in marketing spend is reflected in the evidence on the proportion of the firm's annual budget (or total expenditure) that is spent on marketing. A substantial 42% of respondents spent 5% or less of their annual budget on marketing, 28% spent between 6% and 10%, while a final group of 30% of firms spent 11% or more of their annual budget on marketing. Put rather crudely, one-third of firms spent very little on marketing, while a further third spent a significant amount on this key function. Although there was little sectoral variation in these figures, such figures further support the argument for sharp differences in marketing vigour, suggested in Table 5.12 above, and in further analysis below.

TABLE 5.13 *Incidence of full-time sales staff by percentage employment growth*

	0-25%		25-100%		100%+		Total	
	N	%	N	%	N	%	N	%
Sales staff	18	41.9	29	59.5	36	85.7	79	62.2
None	25	58.1	17	40.5	6	14.3	48	37.8
Total	43	100	42	100	42	100	127	100

Chi Squared = 17.57 P = 0.00015

Clear evidence of a strong commitment to sales and marketing is the existence of full-time sales staff in survey firms. Overall, a substantial 61% of survey firms maintained full-time sales staff. Table 5.13 indicates that, for those firms where employment change data are available, there is a clear relationship between the existence of full-time marketing staff and high growth rates, a phenomenon noted elsewhere (Oakey at al 1988). In terms of employment growth, Table 5.13 clearly indicates that while 42% of firms in the 25% or less growth category maintained a full-time sales force, the equivalent figure for firms in the over 100% growth category was 86%, yielding an extremely strong significance from Chi Squared testing at the $p = 0.0002\%$ level. Other analysis of the presence of a full-time marketing staff with turnover growth, again indicated that there was a strong relationship between these two variables. This previously noted general relationship between marketing investment and growth presents a problem in that, although some form of causal relationship exists, the *direction* of causality is not clear. While it is tempting to suggest that firms with high growth rates are successful *because* they employ a full-time marketing staff, it could also be argued that the success of their product has caused them to employ a full-time sales force. The reality is probably not clear cut, and the most likely explanation for this relationship is that new high-technology firms with highly marketable product ideas begin life with *both* a highly saleable product and a sales force with which it is promoted from an early stage in the life of the firm.

Although there was no detectable difference between the three survey sectors in terms of the existence of a full-time sales staff, there was sectoral variation in terms of the numbers of full-time sales staff employed in the firms of the survey sectors. Table 5.14, which relates the number of sales staff to the survey sectors, clearly indicates that software firms are most strongly represented in the less than five worker category, which is the main reason for a Chi Squared test showing significant difference at the $p = 0.048\%$ level. However, the small number of sales staff in the software firms must largely be explained by the size structure of firms in this sector, since 60% of software firms employed less than ten workers in total. This pattern of results suggests that a majority of survey firms in all sectors maintain a reasonable sales and marketing capacity, while a minority rely on a variety of (possibly inadequate) promotion methods, of which "word of mouth" is especially important.

TABLE 5.14 Number of full-time staff by sector

No. of sales staff	Biotechnology		Electronics		Software		Total	
	N	%	N	%	N	%	N	%
1-4	12	46.2	18	56.3	22	84.6	52	61.9
5-9	10	38.5	11	34.4	2	7.7	23	27.4
10+>	4	15.4	3	9.4	2	7.7	9	10.7
Total	26	100	32	100	26	100	84	100

Chi Squared = 9.59 P = 0.0478

Indeed, Table 5.15, which relates the presence of full-time sales and marketing staff to the main methods of sales promotion, reveals a number of variations of statistical significance through Chi Squared testing at the $p = 0.007\%$ level. In particular, while 42% of firms with no sales staff relied on "word of mouth" as a major promotion method, the equivalent figure for those firms with sales staff was 17%. Conferences and exhibitions were clearly more important for firms possessing sales staff with 23% of these firms acknowledging this method, compared with a modest 6% of firms attending exhibitions that had no full-time sales staff. This result almost certainly stems from the ability of firms with full-time sales personnel to staff stands at exhibitions for a whole week if necessary. However, Table 5.15 also makes it clear that sales visits are not impossible for firms with no full-time sales staff, since visits can be made by senior staff, often the managing director on a part-time basis. In such firms, visits accounted for 17% of main promotion methods, although such contacts are necessarily less frequent than for those firms with a full-time sales staff, where the incidence was 28%. Overall, firms with a full-time sales staff displayed a much wider range of sales promotion methods, when compared with 42% of firms without full-time staff relying on "word of mouth" for their sales.

TABLE 5.15 *Methods of sales promotion by incidence of full-time sales staff*

	Sales staff		None		Total	
	N	%	N	%	N	%
Word of mouth	13	16.5	22	42.3	35	26.7
Adverts	11	13.9	10	19.2	21	16.0
Mail shots	8	10.1	5	9.6	13	9.9
Conferences/ exhibitions	18	22.8	3	5.8	21	16.0
Visits	22	27.8	9	17.3	31	23.7
Other	7	8.9	3	5.8	10	7.6
Total	79	100	52	100	131	100

Chi Squared = 15.93 P = 0.007

TABLE 5.16 *Number of full-time sales staff by sector*

No. of sales staff	<25%		25-100%		100%+>		Total	
	N	%	N	%	N	%	N	%
<5 workers	15	83.3	21	84.0	14	38.9	50	63.3
5 workers +>	3	16.7	4	16.0	22	61.1	29	36.7
Total	18	100	25	100	36	100	79	100

Chi Squared = 16.95 P = 0.0002

Further support to the link between investment in sales and marketing and growth is provided by Table 5.16, where it is clear that there is a link between increasing numbers of sales staff employed and employment growth. Indeed, 61% of the firms in the 100% and over growth category employed five sales staff or more, compared to 39% of firms with less than five workers, sufficient difference to produce a Chi Squared test significant at the p = 0.0002% level. Thus, it is clear that there is a general link between increasing sales staff employment and firm growth. However, as previously noted above, the relationship between marketing investment and the expansion of these new high-technology firms is complex. It is certainly true that vigorous marketing of a poor product is unlikely to be a profitable strategy. Discussion of the complex causality involved will be an important topic for the conclusions to this chapter.

Marketing agreements

In the new firms of this study, where financial resources are both limited and placed under great stress by many competing needs, any strategy towards sales and marketing that can achieve sales at a minimal cost should be attractive. This general principle particularly applies to export marketing (already established as very important in most survey firms), where contact with customers both initially to sell products and during subsequent sales and service arrangements can be expensive if delivered directly by the survey firm. Thus, various forms of marketing agreements *should* be attractive to high-technology small firms, although previous relevant research has found them to be less popular than might have been expected (Oakey 1984; Oakey et al 1988).

Table 5.17 indicates the incidence of marketing agreements in survey firms. Overall, agreements existed in 57% of the survey population. Although differences in the table are not statistically significant, agreements were slightly higher for biotechnology and software firms than was the case for their electronics counterparts. The significant minority of firms with no marketing agreements testify to the reality that, in practice, marketing agreements are more difficult to establish and maintain successfully than they appear in theory. A fundamental problem appears to be that, although such arrangements are cheaper than internal efforts, they are also less effective. Put simply, partners in any marketing agreements are less likely to "push" the product concerned than would be the case if it was promoted by the manufacturing firm itself.

TABLE 5.17　*Incidence of marketing agreements by sector*

	Biotechnology		Electronics		Software		Total	
	N	%	N	%	N	%	N	%
Marketing agreement	28	63.6	20	43.5	30	63.8	78	56.9
None	16	36.4	26	56.5	17	36.2	59	43.1
Total	44	100	46	100	47	100	137	100

Chi Squared = 5.114　P = 0.0775

With regard to the types of marketing arrangement undertaken by survey firms, 43% made use of an agent or distributor, while the other major form of marketing agreement concerned association with another company in 37% of the survey population. The associated firm would include the product of a survey firm in its own product catalogue, and sell (and where appropriate, provide service) in another region, or abroad. While the use of agents or distributors is particularly subject to the above criticism of lack of commitment to selling the client's product, the use of an associated firm is a more permanent, and valuable form of association. Particularly where foreign sales are concerned, if it is remembered that many of the high-technology products of the survey firms may need repair or after sales service, a marketing agreement with a foreign firm possessing similar technological skills, and a relevant sales network, can achieve export market penetration and servicing facilities at a relatively low risk in terms of capital investment, certainly when compared to direct independent exporting efforts.

The competitive environment

A final question in this section on customer relationships asked if market competition had intensified, eased, or remained unchanged between establishment and the time of survey. A clear majority 66% of firms claimed that the competitive environment had intensified since formation, 28% noted that the situation was unchanged, while a small minority of 6% of firms indicated that the position had eased. This small proportion of firms in the "eased" category is almost certainly influenced by the growing recession during the study period. Interestingly, the only variable to be related to this pattern of response that showed any notable result was age of firm. While 77% of firms in the oldest "over nine years old" category stated that the competitive environment had intensified, this compared with 51% of firms that were less than five years old (Table 5.18), sufficient difference to produce a Chi Squared test significant at the $p = 0.02\%$ level. This tendency for older firms to experience marginally greater competition may be an indication of growing sales and marketing problems as the firm becomes established, and matures into a serious competitor for market share in a relevant market niche.

TABLE 5.18 *Change in competitive environment by age*

	0-5 yrs		6-8 yrs		9-13 yrs		Total	
	N	%	N	%	N	%	N	%
Intensified	21	51.2	26	65.0	41	77.4	88	65.7
Eased	6	14.6	1	2.5	1	1.9	8	6.0
Unchanged	14	34.1	13	32.5	11	20.8	38	28.4
Total	41	100	40	100	53	100	134	100

Chi Squared = 11.65 P = 0.020

Summary conclusions

The wide-ranging evidence of this chapter has clearly indicated for all three high-technology sectors of this study that input and output linkages are indeed of great geographical and organisational complexity. Compared with other lower-technology small firms, the non-local supplier and customer relationships in general, and international linkages in particular, have testified to the complex pattern of interactions that sophisticated high-technology production causes such manufacturing firms. However, it is ironic that a generally high level of geographically complex input and output linkages should, within the study, have been distorted by the exceptionally high international distribution of linkages in the biotechnology sector firms. Indeed, although the spatial distribution of electronics and software firms suppliers and customers would have normally been considered impressive, they appear rather parochial when compared with their biotechnology counterparts. The only strong pattern of local relationship concerned the purchasing behaviour of software firms. A suggested explanation for this local orientation was that this represented a very low level of input material purchase overall, since most of the value added in production for software firms was comprised of "intellectual value added", while most input materials were of low technical complexity, and could be purchased locally.

With specific regard to the customers of survey firms, there was found to be an unusually high concentration on both single customer industries, and single customers purchasing over 10% of the survey firms' outputs. This was noted to be higher than for other similar recent studies, and might possibly reflect the difficult trading conditions for survey firms over the study period. Clearly, a wide customer base is not possible if demand is poor and the patronage of single large customers is the only patronage available. In such circumstances, the desire for a balanced portfolio of customers must be sacrificed in favour of secure business. Again, the sectoral data revealed that the biotechnology firms presented a generally extreme position. While biotechnology firms had a strong tendency to maintain a single customer industry and, by 1991, single important customers, the most striking evidence indicated that both industries and customers were predominantly international in their locations. This result confirmed a particularly strong pattern of input *and* output linkages for biotechnology firms in which there was a consistent international orientation.

The results on sales and marketing effort displayed a sharply contrasted pattern of data in which an absence of sales staff and a preference for "word of mouth" and journal advertising methods were popular among a minority of survey firms, while in a majority of cases, a full-time sales staff was employed for which a larger range of sales methods were applicable. There was also a clear link between the extent of investment in sales and marketing and firm growth, although it was strongly asserted that any argument that increased marketing effort *caused* growth would need to be seriously qualified. It was noted that firms with attractive products would be more likely to sell them vigorously, and that no amount of marketing would sell a poor product. It is almost certainly the case that a well conceived new high-technology small firm would be based both on an attractive product with market potential, and on the resources with which adequate marketing effort could be employed.

A final comment on all the data in this chapter might conclude that the networking patterns observed are typically those of specialist "niche" producers. The great strength of "niche" production for high-technology small firms is that most forms of this type of production will not attract the acquisitive attention of large firms since they would not find the relatively low volume of business attractive. Thus such small producers can occupy a lucrative "niche" without any great fear of severe competition. However, there are two major problems associated with "niche" production. First, as implied above, growth within a "niche" can be very slow, which may frustrate ambitious new high-technology small firm owners. The problem for such firms, assuming they can develop a product that has mass market rather than niche potential, is that emergence from the niche will immediately attract the acquisitive attentions of larger firms, which may or may not be welcome.

Second, niche production can become precarious if the niche suddenly suffers a severe recession. The evidence of this chapter has indicated a high degree of sectoral and customer dependence. It is not difficult to imagine, given recent experience in the United Kingdom, a sudden cut back in consumer demand in a given sector, especially when the consumer is the government (e.g. defence; education; health). For example, for a medical instrument producer, a sudden reduction in the purchasing power of United Kingdom health providers could be extremely damaging. While the traditional "fall back" position for such United Kingdom producers has traditionally been to expand exports, a more general world recession since 1989 has often meant that foreign markets are as depressed as domestic sales outlets.

These market problems may be seen against a background of very high operating costs in terms of both procurement and sales (when borne by the producer). While it has been justifiably argued that such costs can be reasonably borne by survey high-technology small firms during periods of prosperity when demand is high, it is also conversely true that such costs render price reductions very difficult to achieve when competition increases due to shrinking market size, as noted by a majority of survey respondents. While the cosmopolitan nature of linkage relationships within the biotechnology firms of this study may be an impressive reflection of their technical sophistication in production and sales, the fact remains, and is clear from Chapter 3, that few of these firms are profitable. The acid test for survival will be, as has always been the case for high-technology small firms, is the basic product technology of the individual firm of sufficient actual (or potential) technical attractiveness to produce sales volumes and unit profit margins that can compensate for otherwise overly expensive production costs? The short-termist attitudes of many external funders of NTBFs do not suggest that long-term views will be taken when assessing the advisability of continued support, either during a short-term recession, or through the substantial periods of development discussed in Chapters 3 and 4 above.

CHAPTER 6

Acquisitions

Introduction

The motivation for the formation of a new high-technology small firm can take many forms. A frequent source of new firm formations in high-technology industries are "spin offs" from existing "incubator" large firms (Speigelman 1964; Freeman 1982; Rothwell and Zegveld 1982). As noted in previous chapters, a major driving force for these entrepreneurial acts is the *autonomy* that is associated with small firm ownership. Indeed, from an innovation viewpoint, the initial product idea for the new firm may be an invention or innovation that the previous large employer firm did not deem economically viable (Oakey 1981). Thus, the benefits of such a culture change for the firm founder in terms of newly won independence, and the job satisfaction that accompanies total control over intellectual property development, tends to make many new firm owners intensely independent in their strategic outlook. In such instances, any form of external control of the business is resisted. This motive explains the prevalent desire for financial autonomy noted in Chapter 3 above. In these circumstances, the sale of equity in return for investment finance, or the acquisition of loans from banks (which might involve a bank member on the board of the recipient firm) are avoided if at all possible.

Conversely there also exists perhaps a smaller group of NTBF owners who adopt a radically different approach. From conception, these new high-technology firms are founded with the sole aim of maximising financial gain for the founder (or founders). Such firms are frequently *founded* on the basis of the substantial external involvement of venture capital organisations. Indeed, for these firms, business plans at the time of formation are principally concerned with attracting substantial external funds, to which the personal capital of the founders may be added. As suggested and confirmed in Chapter 3, many new biotechnology firms begin life with no prospect of an early return from product sales, which ensures that they are particularly dependent on external funding for their initial formation and early growth. In these and other instances where external "start up" funding or ownership is involved, managements of such firms accept the need to sacrifice full internal autonomy, and adopt a management style that is concerned with the persuasion of external investors of the merit of management decisions, rather than acting in a totally independent manner. It has been noted elsewhere, however, that the strain of such investor portfolio management over time can be particularly onerous and debilitating for new biotechnology firms (Oakey et al 1990). There is a danger that incessant consultation with external equity holders can cause excessive bureaucracy and inhibit the firm's management from achieving what it does best, which is the relatively efficient development of high quality technical innovations.

However, although the ownership strategy of the firm might be clear cut in the early years following formation, steady growth brings a number of differing problems that must be tackled by both fiercely independent firms, and those that are eager to "sell out" at the first viable opportunity. In particular, the firm for which independence is highly valued may encounter capital shortages that can result from a number of growth problems. At the two extremes, additional capital might be required, either to allow expansion in increasingly competitive markets as the firm expands, or to buttress the finances of a firm during a period of trading difficulty. Faced with severe problems of this type, independent firms may be forced to seek short-term external financial support as a means of ensuring long-term independent survival. Clearly, in many instances, an envisaged short-term involvement with external investors becomes a "slippery slope" to a total loss of ownership.

None the less, the firm committed to the ultimate selling of the firm's equity also must maintain a difficult balancing act. For many new biotechnology firms, that are not enjoying profitability, the main problem is not "when to sell", but how to keep external investors from *withdrawing* their support in the face of only the *promise* of long-term capital gains through highly profitable product sales. However, new small firms in any high-technology industry may be the possessors of embryonic powerful technologies that have clear long-term value, notwithstanding their poor current performance. In these cases, and in cases where firms have begun to produce modest profits for the first time, the founders will seek to delay the selling of their equity to a point where it reaches its highest possible value. Clearly, the best negotiating position for such a firm is one in which the sale of the firm is conducted from a position of financial strength where *both* the options of selling out and continuing are viable. Many firms with valuable product technologies are forced to sell their technological assets as part of the receivership process because the development of the technology concerned has bankrupted the firm (Oakey 1984). Thus, although a number of new high-technology small firm founders may be prepared to sell out in principle, the timing of the sale is difficult to judge to give optimum value. To sell early may be a financial disaster, while delaying in the hope that a good position will further improve runs the risk of selling "after the peak" which, after all, is only defined by a subsequent slump!

The above comments are all of vital relevance to the subject of acquisition. Given the now well established shortcomings of the United Kingdom capital market as it applies to new high-technology ventures (Murray and Lott 1992; Deakins and Phillpot 1994), the injection of external capital into United Kingdom high-technology firms often takes the form of an exchange of capital for equity. This exchange can take the form of either a minority stake that might be increased to full ownership at a later date, (a practice common in the biotechnology industry [Oakey et al 1990]), or a full and final purchase of the high-technology small firm concerned.

In cases where the small firm involved judges that the sale is welcome in the context of the above discussion, the acquisition may be mutually beneficial. However, in cases where the acquisition is prompted by a need for investment capital, the decision to sell equity on the part of the small firm ownership is often driven by *financial necessity*, and not by a desire to realise assets at an optimal moment.

While it can be argued that the acquisition of small firms by their larger counterparts in the same (or related) industries is merely part of the competitive process in a market economy, it is also conversely true that the electronics industry has benefited, in industrial innovation terms, from the contribution of a healthy *independent* small firms sector. If the small firms of an industry are acquired by large firms in the early stages of the industry life cycle, there is a danger that diversity of technological approach will be lost, and that the acquisition of small firms by their larger counterparts might be a "Luddite" strategy with the express intent of gaining control of competing technological approaches to problem solving (Oakey 1993). Since a minority of the survey firms of this study have been acquired during the study period, there is some opportunity to discover what influence acquisition has had on these new enterprises.

The incidence of acquisition in survey firms

Initially, it is of interest to discover the extent of acquisition in the survey population, and any particular characteristic of the acquired firms. An obvious general trend in the data concerns the timing of acquisitions. While 26 firms (19%) had been acquired by 1991, 20 of these acquisition had occurred since 1986, during the second half of the survey period. While a number of the acquiring firms may have held a stake in these 26 firms prior to acquisition (and indeed, perhaps from formation), the most likely explanation for the more recent increase in acquisition activity in survey firms is the need for a "track record" of performance to emerge on which a case for full acquisition by the acquirer could be based.

TABLE 6.1 *Ownership status of firm by employment*

Workers	Acquired		Independent		Total	
	N	%	N	%	N	%
<10	1	3.8	45	40.5	46	33.6
10-49	13	50.0	53	47.7	66	48.2
50+>	12	46.2	13	11.7	25	18.2
Total	26	100	111	100	137	100

Chi Squared = 22.16 P = 0.00002

There is a strong relationship between acquisition and high growth performance in acquired survey firms. This assertion is generally supported by Table 6.1 which confirms that it is the larger firms of more than ten workers in general, and 50 workers in particular, that are the main targets for acquisition. While only 2% of acquired firms fell in the less than 10 employees category, 48% of these purchased firms employed over 50 workers. This association proved *strongly* significant during Chi Squared testing at the p = 0.00002% level. Here the main stimulus for acquisition would appear to be the existence of a "track record" of performance *over time* (noted above), since there was also a statistically significant association at the p = 0.016% level between the propensity for acquisition and age of firm (Table 6.2). While a small 8% of acquired firms had been purchased within five years of formation, this percentage rose in the 9-13 year category to 58%. A "track record", represented by the linked attributes of age and size, appears to be a major trigger to acquisition. The *early* full purchase of small new firms does not appear to be a popular strategy.

TABLE 6.2   *Ownership status of firm by age*

Age	Acquired		Independent		Total	
	N	%	N	%	N	%
0-5 years	2	7.7	39	35.8	41	30.4
6-8 years	9	34.6	31	28.4	40	29.6
9-13 years	15	57.7	39	35.8	54	40.0
Total	26	100	109	100	135	100

Chi Squared = 8.24 P = 0.016

Although marginally outside the normally accepted level of statistical significance accepted for Chi Squared testing at p = 0.063%, it is clear that acquisition is lower in the software sub-set of survey firms at 9%, when compared with 27% for biotechnology firms and 22% for electronics enterprises (Table 6.3). It has been noted in Chapter 3 that software firms experience both low barriers to entry and low subsequent growth rates. Indeed, while more software enterprises fall in the less than ten worker employment size category within the survey total, they are *not* significantly younger than their biotechnology or electronics counterparts. Thus, since software firms are not achievers of fast growth, they may not be attractive acquisition targets for investors in high-technology industry.

TABLE 6.3   *Ownership status of firm by sector*

	Biotechnology		Electronics		Software		Total	
	N	%	N	%	N	%	N	%
Acquired	12	27.3	10	21.7	4	8.5	26	19.0
Independent	32	72.7	36	78.3	43	91.7	111	81.0
Total	44	100	46	100	47	100	137	100

Chi Squared = 5.54 P = 0.0625

Table 6.4 confirms the link between firm growth and acquisition by indicating that the majority of acquired firms fell into the high employment growth category. While 58% of acquired firms experienced employment growth of 100% or more, the equivalent figure for independent firms was 27%. Although it *might* be argued that such generally high growth in acquired firms was an *effect* of acquisition and not its cause, previous evidence that acquisitions have mainly been recent events suggests that there has not been enough time for "post-acquisition" impact to have caused the relationship in Table 6.4. It is far more likely that the acquirers were attracted to firms with *previously impressive* independent growth.

TABLE 6.4 *Ownership status of firm by percentage employment growth (1987-91)*

%	Acquired		Independent		Total	
	N	%	N	%	N	%
0-25	5	20.8	38	36.9	42	33.9
25-100	5	20.8	37	35.9	43	33.1
100+	14	58.3	28	27.2	42	33.1
Total	24	100	95	100	127	100

Chi Squared = 8.53 P = 0.014

Acquisition criteria

The motivation for acquisition

Since a minority 26 firms within the survey have been acquired, there is little value in attempting to produce statistically significant contingency tables with regard to analysis of the factors that might be associated with the acquisition of these enterprises. Rather, key factors surrounding the acquisition will be discussed in the text below, with simple supporting percentages where appropriate. However, although the acquired firms are a minority of the total survey population, the importance of acquisition is enhanced if it is recalled that the tendency is *increasing* (Oakey at al 1990), and that full acquisition may be the final state of a process that begins with partial ownership and previously unsuccessful acquisition attempts, phenomena noted to be common in the independent firms discussed below. With regard to the main reasons for "selling out" on the part of the previously independent survey firm managements of the 26 firms concerned, 17 (65%), accepted the acquisition due to a strong need for capital investment. These owners can be distinguished from a smaller group of three owners (12%), who *intentionally* sold the business in order to realise their original investment. This is an important distinction, since it implies that a majority of the acquired firms were not voluntarily acquired in that, had other capital investment options been available, at least some of the 17 firms accepting acquisition for financial reasons would have chosen to remain independent.

When asked why the acquirer purchased the firm concerned, two major interesting reasons were given. Twelve of the respondents (46%), cited the acquisition of the product technology of the acquired firm as the main reason for the purchase. A further nine firm respondents (35%), claimed that the main reason for the acquisition was the desire of the acquirer to make a strategic investment in a new sector, by diversifying away from their main area of business. Interestingly, in their different ways, both these tactics are a compliment to the ingenuity of new high-technology small firms, and suggest the good value for money their technical assets can become, whether accessed to add breadth to technological expertise in the same product area as the acquirer, or to allow expansion into a totally new area of production.

Previous association with acquirer

In terms of geographical relatedness, 13 (50%) of the acquiring firms were located in the United Kingdom, 6 (23%) emanated from Europe (5 from the EC), and 5 from North America. With regard to sectoral relatedness, 15 (58%) of the acquirers were from the same or related technical field as the acquired firms, while a further 10 acquirers (38%) were from unrelated sectors of industry. These data concur with the above information on reasons for acquisition, since almost equal proportions of acquiring firms were either seeking to expand within the sector concerned, or intending to expand into the acquired firm's technological area of business.

When asked if the acquired firm had maintained a link with the acquirer prior to the acquisition, 11 firms (42%) acknowledged such a prior link. These affirmatively responding interviewees were asked to indicate further the nature of the acknowledged link. While 5 respondents (19%) stated that the link was a customer or supplier trading link, a further 5 firms noted that the link was a previous minority share holding in the firms prior to acquisition. This evidence supports earlier work by indicating that previous trading or part ownership links are a common precursor to acquisition by the large firms concerned. Such "privileged" contacts allow the large firm to keep a "watching brief" on the growing small firm that provides key "inside information" on progress which subsequently enables the acquisition to occur at a time most beneficial to the acquirer (Oakey et al 1990).

Given earlier work on acquisitions in high-technology industry (Oakey 1981), it was not surprising to discover that a strong level of post-acquisition control was allowed to the acquired firm by the acquirer. In 16 cases (62%), the level of post-acquisition control in the acquired firm remained total, although in a further 5 cases (19%), autonomy remaining in the acquired firm was partial, while in only 5 cases (19%) did independent control totally pass to the acquiring firm. Moreover in only 2 cases was it acknowledged that the new post-acquisition control arrangements had inhibited development of the acquired firm.

The value of the acquisition to acquired firms

Finally, the survey examined the main advantages and disadvantages of the acquisition, and the nature of the "ongoing" relationship. The main advantage was clearly the injection of cash in 13 cases (50%), while "a fresh approach" was cited in 4 instances (15%), and rather dramatically, "survival" was the acknowledgement of 2 firms (8%)! Conversely, perhaps for internal political reasons, the disadvantages were less frequently acknowledged. While 4 respondents (15%) cited loss of control, and 3 firm executives (12%) noted increased bureaucracy, the remaining responses were not consistent in terms of the problems cited. This generally favourable impression of the relationship between acquirer and acquiree is further enhanced by the observation that 10 (38%) of acquired firm respondents claimed that their relationship with their parent was "good", and a further 4 respondents (15%) asserted that the relationship was "hands off", leaving them full autonomy. Only one respondent noted that the relationship with the parent firm was bad. Indeed, it is likely that the strong importance of funding in the set of reasons for the acquisition may partly explain the enthusiasm for the acquisition in acquired firms. Compared to struggling on in difficult financial circumstances (or closure), the involvement of an external partner, given that autonomy was not an option, might be a desirable "second best" solution.

The strategic relevance of acquisition to survey firms that remain independent

While the issue of acquisition is of central importance to the above firms that have been acquired by other mainly larger firms, the issue of acquisition is a constant concern for many more independent firms in the survey sample. A minority of firms may actively seek to purchase other mainly small firms as part of a rapid growth strategy. In other more frequent instances, the possibility of "selling out" all or part of the firm's equity is a constant strategic option for independent survey firms, that is often triggered by offers of purchase from potential large firm acquirers. As noted in the introduction to this chapter and other parts of this book, given the generally poor capital market for small firm investment, the sale of a company is a major and recurrent potential means of acquiring often much needed external capital investment.

Attempts to acquire other firms

As might be expected given the youth of the survey firms, and their consequently relatively meagre financial resources, the number of firms that had acquired other firms since formation was small. A modest 15 firms had acquired businesses since birth and, perhaps significantly, it is interesting to observe that 12 of these firms were engaged in electronics production. While 3 acquiring firms were derived from the biotechnology study sector, *none* of the software survey firms had acquired another firm. In keeping with a number of other results in this chapter on acquisition, the incidence of acquisition *by* survey firms appears to be linked to rapid growth and a consequent larger than average size of the acquiring survey firm. All acquiring firms employed more than 10 workers, 13 of these enterprises fell into the over 25% employment growth category, while 8 of these enterprises recorded employment growth rates of over 100%. The absence of software firms from this small group of acquiring firms must partly be explained by the previously noted slow growth and small size characteristics of this sectoral sub-group. The comparatively high level of acquisition activity in the electronics sub-group must also be, to some extent, related to the strong production orientation of this sector, and the scope for synergies and economies of scale in amalgamating complementary forms of manufacture (e.g. the purchase of a printed circuit board maker by a component assembly firm).

Unsuccessful acquisition attempts towards independent survey firms

Independent firms often receive welcome or unwelcome approaches by other firms seeking to make a full or partial purchase of the target firm's equity. As discussed above, the degree to which the reaction of survey firms is favourable depends on *both* the overall long-term objectives of the firms concerned, and the appropriateness of the timing. While many firms would always view a successful acquisition attempt and the associated loss of control as a "fate worse than death", other firms with the long-term strategy objective of "selling out" might refuse to sell on a given occasion, as argued above, only due to the inappropriateness of the timing of the offer, or other conditions of the proposed sale (e.g. a poor proposed sale price).

The assertion that acquisition attempts towards NTBFs are frequent is confirmed by the initial observation that 48 of the independent firm sub-sample (i.e. 44%) had been subjected to a serious acquisition attempt. If these firms are added to the 26 firms that had been fully acquired during the study period, a majority group of 74 firms is derived that had been subjected to strong acquisition activity (i.e. 54% of all survey firms). Interestingly, there was no significant sectoral bias in these data, with the level of acquisition attempts in each of the three study sectors not deviating substantially from the total independent firm average. However, a growing body of evidence of a link between acquisition interest and growth of target firms is confirmed by Table 6.5, where it is clear that there is an association between the propensity for acquisition attempts and employment growth in survey firms in the instances where turnover growth data are available.

TABLE 6.5 *Incidence of acquisition attempts by percentage turnover growth (1987–91)*

	Acquisition		Attempted		Total	
	N	%	N	%	N	%
0–24	5	12.2	14	28.6	19	21.1
25–100	12	29.3	19	38.8	31	34.4
100+>	24	58.5	16	32.7	40	44.4
Total	41	100	49	100	90	100

Chi Squared = 6.78 P = 0.033

While the rate of acquisition attempts in firms experiencing employment growth of 25% or less was 12%, the comparative proportion for the 100% and over employment growth category was 59%, a difference sufficient to produce a Chi Squared test significant at the p = 0.033% level. When questioned on the reasons for the failure of the acquisition attempt noted by survey firm respondents, the responses support strategic behaviour noted to be important earlier in this chapter. A strongly independent group of 18 survey firm respondents (38%) stated that their independence was highly valued, and this would have been lost if acquisition took place, while a smaller group of 12 respondents (26%) claimed that, although they were prepared to sell, the offer was either ill-timed or not large enough.

These two "real world" attitudes exemplify the alternative strongly differing management approaches of those firm owners who resist acquisition, virtually at any price, and those who strategically would be prepared to sell "when the time is right". Indeed, a small minority of 13 survey firms pro-actively sought to sell their firms to an external buyer. The sectoral pattern of results interestingly complement earlier evidence on acquisition attempts in that the strong showing by survey electronics firms in terms of acquisition *was not* reflected in any attempts to sell themselves, since only one electronics firm had attempted to "sell out". Conversely, while it may be remembered that none of the software firms in the survey had sought to acquire other firms, six of the survey firms in this sub-group did *try* to sell their company. The remaining six firms seeking to sell their businesses were biotechnology companies.

Although it is always dangerous to generalise, it is difficult to escape the conclusion that firms seeking to purchase other companies are acting from a position of strategic strength, while conversely, firms that attempt to sell themselves are often in severe strategic and financial difficulties. This assertion, taken together with other performance evidence from Chapter 3, suggests that the electronics and software firms represent two extremes in which the electronics firms are characterised by fast growth, larger size and aggressive strategic behaviour towards acquisition, while software firms tend to be small, slow growing and, perhaps not surprisingly as a result, comparatively unattractive as acquisition targets, and do not provide a strong financial basis upon which acquisition attempts might be founded.

The envisaged strengths and weaknesses of acquisition to independent firms

As a final question to these independent firms on acquisition, respondents were asked to speculate what would be the major strengths and weaknesses of acquisition, should it occur. A number of rather reassuring responses were obtained from this questioning that support earlier results. In terms of the potential benefits of acquisition, a large group of 43 firms (40%) cited capital investment as the major advantage of acquisition. It is likely that this strong minority view reflects the reality that, given a very poor capital market for independent firms in Britain, acquisition remains a major alternative source of capital funding. However, in view of the strong sense of independence felt by many new high-technology small firms, this option is often not an attractive solution. A further 11 survey firms (10%) noted that personal gain would be a major attraction of acquisition, again reflecting the converse "grow to sell" strategic approach.

In terms of disadvantages, by far the most important reason for resisting acquisition was a perceived "loss of control" in 82 survey firms (76%). No other cited disadvantage was mentioned by any substantial numbers of firms. This result reflects the clear reality that independent control is a major part of the "psychic income" of new high-technology small firm founders. While, in desperate financial circumstances, survey firms' owners may be forced into selling their business, the founders that form a business with the express objective of "selling out" at the most auspicious time are comparatively rare. The clear evidence indicating the need for capital as a major reason for sale of acquired firms, together with this other evidence on the desire for autonomy, suggests that, for many firms, the agreement to acquisition is born of necessity rather than design.

Summary conclusions

The threat or promise of acquisition is a constant issue confronting most new high-technology small firms. Except in instances where poor performance in terms of innovation and growth renders such firms unattractive to potential acquirers, acquisition remains an issue for all survey firm managements. As discussed above, the acceptability of "selling out" depends on a whole range of factors associated with the long- and short-term objectives of the firm, and the external economic pressures that may cause fiercely independent firms to consider selling all or part of the firm's equity to gain a return on their emotional and financial investment.

However, for strongly independent new high-technology small firms, based on close internal team work that is the essence of small firm innovative success (Rothwell and Zegveld 1982; 1985), acquisition may not only be difficult to accommodate in terms of the autonomy of the firm. Loss of control may, perversely, severely damage the very success that made the firm an acquisition target at the outset. There is a sense in which new high-technology small firms are not best served by absorption into large bureaucratic organisations. This reality is partly reflected in the willingness of the new owners of acquired firms (noted in the analysis of this chapter) to allow substantial autonomy in acquired firms following full purchase. None the less, such autonomy, based on the discretion of others, only partly compensates for the full intellectual property ownership that is only complete (and completely motivating) when the firm is owned and controlled by the original founders. In truth, for many new high-technology small firm technical entrepreneurs, the worth of owning a firm is comprised of a confused mixture of intellectual pride and financial self-interest in which the balance between these two features will vary depending on the prosperity of the firm and the personality of the founder.

While it is clear that a majority of new high-technology firm owners would prefer to remain independent, the need for a financial return on their investment (particularly if their house is second mortgaged against the enterprise) will become of critical importance at times of financial stress. Given the previously noted poor financial capital market for small firms in the United Kingdom, it might be prudent to abandon a desire for freedom of action, in favour of acquisition, if acquisition is the only means of gaining *some* recognition for the work put into a business. It is a perversity that acquisition may be the only viable option for successful fast growing firms in need of substantial injections of investment finance. This has been a reality that has existed for high-technology small firms throughout the 1980s (Oakey 1981; 1984; Oakey et al 1988).

Whether acquisition is a beneficial phenomenon in terms of sectoral innovation and economic development is not clear. While it can be argued that acquisition is merely a result of competitive market forces in which the strength of large firms is merely made available to their growing small firms counterparts, it is also possible that acquisitions tend to stifle radical innovation. It might be argued that innovation in acquired new high-technology small firms is not as radical or effective in a large organisation than would have been the case in a fiercely independent NTBF. It is certainly true that much of the radical and important innovation in the United States semiconductor industry took place in small independent firms that helped develop the industry for all scales of production. Conversely, in the United Kingdom, where much early semiconductor development occurred in large firms, development was slow, and eventually abandoned. There is a sense in which new high-technology small firms do not allow a large firm dominated technological paradigm to become established since, by virtue of their novelty, they have no strategic interest in maintaining the status quo; indeed new small firms with radical new product ideas have a direct interest in destroying intellectual "log jams". Since the acquisition of high-technology small firms by their larger counterparts is a growing phenomenon, a strong case can be made, based on data in this chapter and other research, for a fuller investigation of the overall impact of such full acquisitions on the health of the sectoral, regional and national economies in which they occur.

CHAPTER 7

Conclusions

This book has sought to investigate in detail possible variations in the formation and growth characteristics of the new firms of three different high-technology sectors. A major aim of the study was to test two main hypotheses. First, that *strong* variations in the characteristics of high-technology sectors within a general high-technology industry definition might render the use of such an all embracing categorisation a hindrance rather than an aid to the development of policy and theory regarding such firms. Second, that the nature of the founding technology chosen by the entrepreneurs of high-technology industries will have a profound determining influence on their subsequent strategic freedom. Indeed, while there will always be some scope for quality of management performance in determining success, a major influence on the speed with which progress can be made is the basic product technology of the founding firm that *broadly* dictates the early funding requirements of the embryonic business. The following conclusions are presented in three parts, beginning with a brief unifying summary of the individual chapter findings, followed by a consideration of the implications of these findings for theory, and ending with a discussion of what implications the results of this work have for industrial policy towards NTBFs.

A summary of results

Following the setting out of the main hypotheses of the book in the early part of Chapter 2 on the degree to which founding entrepreneurs are determined in their actions by their chosen product technologies, the second part of this chapter sought to test whether origins played a major part in influencing the chosen business of survey participants. As anticipated, strong evidence was gathered on a link between the "choice" of founding technology and the previous industrial or public sector employer of the main founder of the firm. It was noted, in such cases, that founding "choice" is limited by previous experience, and that expertise developed over the previous years, prior to formation, have a strong deterministic influence on the range of activities "chosen" by most new firm founders.

Chapter 3, which provided detailed financial information on the founding and subsequent early growth behaviour of new high-technology firms, generally confirmed that the founding of a new firm in all the study sectors can be fraught with difficulties. However, perhaps the most striking result to emerge from the analysis of this chapter was the extremely severe difficulties experienced by the biotechnology sub-group of firms. The problem of long lead times associated with product development caused a double problem in which high R&D costs associated with product development were not mitigated by compensating profits from product sales. In the short to medium term, the electronics firms of the survey generally appeared to produce the best performance in terms of early growth following formation. Indeed, although software firms enjoyed the benefits of relatively easy entry costs and a well established core technology, the intense competition that is a consequence of low entry barriers tended to ensure that most software firms were not able to achieve fast subsequent growth.

None the less, it was concluded that the current problems of the biotechnology firms of the survey might be overcome in the long-term, and it was further noted that many of these currently struggling firms may have the best long-term growth potential, provided that they can survive short-term funding problems (further discussed in the policy sub-section below). A strong general tendency in the results to avoid external financial involvement in the funding of the firm, noted in other work (Oakey 1984), was concluded to be partly a result of the onerous conditions associated with such involvement, rather than any absence of a need for external financial support for internal funding.

The findings of Chapter 4, which investigated the impact of R&D activity on the organisation of the firm, augmented the results of Chapter 3, by identifying the main reason for the financial problems of many survey firms. Indeed, the *generally* high levels of R&D investment present in high-technology new firms of the study were exceeded by a number of *particular* biotechnology firms. In keeping with a number of other results of this book, the funding of R&D in biotechnology firms was *extremely* high, often as discussed above in conditions where no income from sales was compensating for such extreme "front end" investment. The R&D investment required in these biotechnology firms tends to confirm further the deterministic element in founding technologies of this type by implying that, while in certain circumstances superior management techniques may bring success to a newly formed enterprise, the embryonic nature of many of the founding technologies of new biotechnology firms ensures that long-term funding is essential to ensure the continued existence of such enterprises.

The material input and output relationships of survey firms, examined in Chapter 5, displayed the expected sophisticated national and international linkage patterns. Again the, by now, established pattern of extreme results from the biotechnology firms of the study were augmented by the results on both input and output linkages. Biotechnology firms maintained linkages that were predominantly international in geographical distribution. However, it was prudently remarked that, although impressive in a technical sense, in many ways these complex and expensive networks represented another source of added operating expense for firms that, in most cases, were not profitable.

A further general trend was the tendency for survey firms to be linked with a single major customer industry and single major customer firm. Clearly, there is interaction here in that, in a number of cases, a single major customer firm will dictate that there is a single major customer industry. In the past, internationally diverse sales linkages have been an asset in that domestic recessions did not severely impact on high-technology small firms (Oakey 1981). However, international recession (particularly in Europe and North America) suggests that survey firms may be vulnerably dependent on narrow market niches. This observation is particularly relevant to the biotechnology firms of the study. They are often forced into seeking international customers due to the highly specialist nature of their products and services where sufficient customers can only be obtained from a world market place.

The final empirical Chapter 6 dealt with the increasingly important issue of acquisition in NTBFs. It was noted that acquisition was of obvious central importance to the recently acquired firms of the survey, but also a continuing strategic issue for those firms that remained independent. While the impact of acquisition on the innovation and growth of new high-technology small firms is not known, it was clear that the sale of survey firms is, in most cases, not the first option that firms in need of external investment would have chosen, given a range of funding options. In this sense, the growth of acquisitions is a further indirect indication of the paucity of external funding for new high-technology firms that does not involve a substantial loss of internal autonomy. Indeed, in many cases, there is no British alternative to acquisition for such firms in need of substantial external investment.

Implications for theory

It is clear from all the results contained in the above chapters that the three high-technology sectors studied display strong differences in terms of formation and early growth characteristics. While it is readily clear that such differences will have strong relevance to the consideration of policy to follow, they also influence theoretical considerations under this heading. Throughout the 1980s there have been a number of attempts to define high-technology industry by establishing criteria that may identify members of a high-technology family of industrial sectors (Breheny and McQuaid 1987; Kelly 1986; Premus 1982; Glasmeier 1985; Markusen et al 1986). Apart from the work of Kelly, who attempted to construct an index of innovation outputs as a measure of high-technology status, most of the other attempts at defining high-technology activities used R&D inputs measures (e.g. either R&D spend or R&D workers). Apart from the problems associated with the use of innovation input measures that can be heavily influenced by defence spending in many western countries where R&D inputs are used to generate military output that often have little commercial value, the rigid criteria used often result in the classification of industries as high technology that are of dubious high-technology pedigree (e.g. soap production [Markusen et al 1986]). Moreover, the discovery that there are significant differences between the study sectors tends to question the term "high technology" as a useful means of defining aggregated high-technology-based forms of production. It is a basic principle of all inductive forms of scientific study that researchers should seek to collect various phenomena and gather them into families or classes as a basis for theory building.

It has been previously argued that any rigid definition designed to assist in our understanding of high technology industry may be dangerous if the definitional criteria proposed are spurious (Oakey et al 1988). In such circumstances, industries may be included as high technology when they are not suitable for inclusion, or even worse, excluded when they should have been included. None the less, a further definitional problem, that has been a major focus for this book, occurs where sectors are *accurately* classified as high technology at an aggregate level, based on a given set of rather crude criteria. This problem is illustrated by the observation that, judged in terms of the input measures used in the above mentioned studies, all the three sectors used in this study would have been easily classified as high technology. However, the above investigations of this book have shown that strong diversity exists *between* these supposedly similar high-technology sectors. Put simply, the problem of a rather crude definition of a phenomenon is that a high degree of spurious homogeneity is assumed for the defined category. This error is particularly severe when the point of departure of theoreticians or policy makers is to assume wrongly that, say, policy towards new biotechnology firms can be identical to policy directed at software firms *because* they are both high technology. Seen in terms of these arguments, the acceptability of a definition must largely be judged in a functional manner in terms of whether the working definition aids or obscures our understanding of the subject of study (Oakey et al 1988). The following explanation of high-technology, based on evidence from the above study, will seek to improve our understanding of this term by dividing high-technology industrial activity into a three-fold functional set of sub-categories.

Most existing definitions of high-technology industry depend on a minimum threshold being met in order for a given sector to be deemed high technology. As noted above, such a threshold is often judged in terms of R&D workers, R&D investment or an index that is a combination of a range of R&D input measures. It has also been observed above that all three industries in this study would easily achieve this threshold and qualify for a high-technology definition. The problem for this study, and all those concerned with high-technology industrial growth, is that such a definition is far too crude to be of use since, as this book has confirmed, there are severe "within category" variations *between* included industries. The observations of the above chapters prompt an explanation for this variability that is based on the core technologies utilised by individual sectors.

Figure 7.1 presents a generalised model of high-technology production in which a development cycle is depicted, ranging from "pre-production", through "launch transitional" to "production". In this case, and in view of evidence from the above study, the three study sectors have been allocated to these three phases to illustrate the argument, by giving "real world" examples of the types of production that might exist within these stages of technological development. Unlike the models presented in Chapter 3, Figure 7.1 should not be viewed as a product life cycle model, judged in terms of the evolution of a single product. Rather, this model should be used to interpret the positioning of a founding product around a "break even" point for a given technology, before which there are a notional ten years of pre "break even" production development, after which there are (at least) ten years of production.

Figure 7.1 Model of general sectoral evolution showing R&D and retained profit evolution

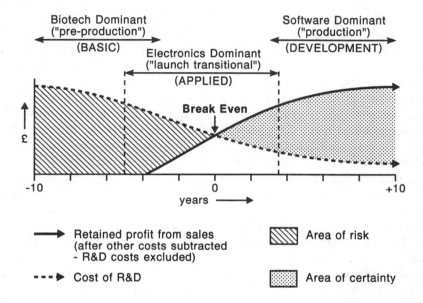

The main value of the model is its reflection of the real degree of "choice" that is available to a founding entrepreneur. As discussed above, depending on his experience, the founder of a new firm will generally base his new high-technology business on his previous experience. While, all things being equal, founders would prefer to begin a new business based on proven technology for which most of the R&D had been performed, and for which a large market had been established, this option is often prohibited by the type of knowledge possessed by the founder. It is the highly generalised argument of the model in Figure 7.1 that biotechnology founders will tend to begin their new enterprises in the pre-production phase, electronics founders will "start up" in the "launch transitional phase" and software firms will enter the technologically relatively established "production" stage of technological development in terms of their founding technologies. This assertion reflects the reality that the "lead times" for the products of the three study sectors have been shown to vary widely where, at the extremes, it is common for a biotechnology firm to endure five to eight years pre-production development of a new product, during which time sales are meagre or non-existent, while software firms can begin production, virtually on the day of formation, and achieve sales shortly after.

The model also includes a consideration of the impact of changing technological uncertainty that accompanies all decisions to begin production which is shown as a "trade off" between the level of R&D expenditure required and the retained profits of the company concerned. In high-technology industry, where R&D costs are such a large proportion of total operating costs, this reality is reflected in Figure 7.1 where there is a ten-year gap between the formation date for the most extreme founding case and "break even", and indeed, a period of four years when no compensating sales are achieved. This extreme proposition has been a reality for a number of biotechnology new firm formations investigated in this book and, as previously discussed in connection with Figure 3.4 in Chapter 3, the long lead times between formation and "break even" might be over-optimistically reduced in the business plans of such firms as a conscious or sub-conscious strategy to gain support from "short-term" oriented external funders.

The model also caters for the often neglected fact that much technological innovation in high-technology sectors is of the "me too" variety, involving various forms of technological copying, after a new technology has reached a proven "break even" point, and also at a stage when R&D costs are no longer prohibitively risky. Indeed, a major reason for the use of the software sector as an example of the "production" phase in Figure 7.1 is that the basic technology of software programming is a commonly known technique, for which the R&D costs have been "written off" long ago, and to which all potential entrants to the sector have access.

However, the previously noted low barriers to entry in the software industry are a mixed blessing in the sense that the increasing levels of technological certainty depicted in Figure 7.1 for the "production phase" of the model are not necessarily a guarantee of financial success. It has been asserted in Chapter 3 that, although the barriers to entry in the software sector are low, many subsequent entrants to the sector, based on a commonly known technology, tend to depress prices that can be charged due to intense competition by many producers for a finite market. This argument is certainly supported by the poor performance of software firms in terms of profitability and growth (see Chapter 3). Indeed, it is again worth emphasising that the model in Figure 7.1 is *generalised* in order to typify a general process of technology development within a sector. In this context, the area of profit or "certainty" indicated in the model is the volume of profit for the whole industry. Interestingly, the extent to which an NTBF can capture part of this business is determined by the level of intellectual property rights (IPR) such a firm can command. In the case of software firms, for example, the gainable proportion of sectoral profits in Figure 7.1, it is argued, will be meagre due to their inability to achieve exclusive control over well established and generally available technology. Conversely, the potential for biotechnology firms to gain exclusive rights over their leading edge discoveries is high, although, as mentioned earlier, the risk of failure is proportionally severe.

This observation leads to a possible general explanation of survey results that might account for the overall growth performances of the three study sectors, expressed in terms of Figure 7.1. The difficulties encountered by new biotechnology firms of the study can be largely attributed to their general formation in the "pre-production" stages of Figure 7.1, where R&D costs are extremely high, compensating sales from production have not yet occurred, and there is no short-term prospect of high profitability. Indeed, Chapter 3 has argued that many of the biotechnologies concerned may have been transferred into new firms at an earlier stage than would be financially prudent, mainly for reasons of intellectual property appropriation. For all these reasons, the short-term growth performance of the new biotechnology firms in the survey has not been impressive, although it was noted that, in the long term, some of these firms may have considerable growth potential. Interestingly, the relatively poor performance of the software firms is attributable to an exactly opposite reason. While the biotechnology firms struggle to survive because they were formed in the "pre-production" zone, too far in advance of the point of "break even" (implying high R&D costs and risks), the software firms were established too far past the point of "break even" in the "production" zone, where the technology is established and generally available, with competition consequently intense.

It could be argued that the reason for the relative success of the electronics firms of this study is their tendency to occupy the "launch transitional" stage of technological innovation around the "break even" point, where the application of technologies offers a "best of both worlds" solution to the foundation problems of NTBFs. Put simply, electronics firms rarely require the extremely long lead times frequent for new biotechnology enterprises, since they either concentrate on the development of new products that require less than five years development, or capitalise on technologies that have recently passed the "break even" point of Figure 7.1. These are the types of technological innovations that, when included in an NTBF business plan, would be generally acceptable to most current short-term oriented venture capitalists. However, it is significant that the technology would also be new enough to protect it from the "me too" follower problems encountered by NTBF software firms. Indeed, if a new discovery occurs in a "niche" area of production, common in the electronics industry, respectable profits might be obtained for a protracted period. Thus, the electronics firms of this study appear to occupy an optimal transitional area in terms of technological development that is neither too risky (as in the case of biotechnology) nor too well established and consequently subject to intense price competition (as in the case of software).

All the evidence of this book in general, and the argument of this chapter in particular, suggests that the basic characteristics of industrial sectors deterministically influence the ease with which new firm founders might begin and develop their businesses. It is also clear that an expert software engineer would find it almost impossible to found a new business in the electronics industry, even if it is accepted that progress in this particular sector might be more easily achieved. Moreover, the argument of this chapter is not "drifting" towards advocating that only new firm formations in the electronics industry are supportable in terms of industrial policy, considered in depth below. The nub of the argument here, based on the accumulation of evidence in this book, is that sectoral technologies *do* impose varying levels of constraint on the potential for new firm formation and growth, that can only modestly be mitigated by superior management techniques. This assertion re-evokes the debate between Woodward and Childs in Chapter 2 on the extent to which technology determines productive efficiency, or the degree to which technology is merely subservient to management techniques, that are argued by many to be the real key to success.

The overall conclusion that might be drawn from all these deliberations is that the term high-technology industry is a term that is currently often pitched at too high a level of aggregation for it to be useful, in terms of either theory or policy making. As a term for the layman it probably has some value, but for the academic researcher the term implies a spurious level of homogeneity that is not apparent when detailed investigations of the type performed in this research are conducted. As Figure 7.1 suggests, high-technology small firms exist along a technological continuum ranging from NTBFs concerned with "basic" leading edge research, to those performing "me too" functions that are often sub-contract in nature. These differences are important, not only in order that we should develop an accurate understanding of *how* high-technology firms function from a theory-building viewpoint, in a rather voyeuristic fashion, but possibly of more importance, in order that we can cater for the varying needs of these new firms, as part of an attempt to aid their growth. The discovery that formation and growth is less problematic in the electronics group of survey firms is not an indirect argument for the abandonment of the new firms of the software and biotechnology sectors. Software and biotechnology engineering are obvious areas where any modern economy must develop capacity into the next millennium, where new small firms will play an important role. From an initial identification of how the high-technology firms of given sectors differ, we must progress to tailor policies to nurture differentially their varying portfolio of needs.

Implications for policy

Although the new small enterprises that emanate from emergent high-technology industries have many potentials, by the end of the 1980s their image had been damaged by ill-informed and grossly unrealistic expectations of both experts and laymen. As discussed in the introduction to this book, initially NTBFs were seen as a panacea for all our industrial growth problems at the beginning of the 1980s, only to be followed by severe disenchantment, witnessed by the flight of venture capital from the funding of this form of industry by the early 1990s (Murray and Lott 1992; Deakins and Philpott 1994). In reality, NTBFs are neither potential saviours of a nation's industrial heritage, nor an insignificant corner of industry worthy of neglect.

Variable sectoral activity: implications for policy

While most NTBFs are not particularly fast growing (Oakey 1991), rarely represent vehicles for a fast return on external investment, and are often high risk investments due to the unproved nature of many of their core product technologies, they none the less deserve continued support for a number of reasons. First, the slow growing, often "niche" producers, within high-technology industries, provide in aggregate secure employment for many thousands of workers in developed western economies. Moreover, their outputs, although not necessarily "leading edge" in technological terms, form part of a national high-technology industrial base that permits the existence of other larger producers in the economy (e.g. in aerospace). Second, there also exists a sub-group of relatively elite new small firms within any high-technology sector that have the potential to revolutionise the industry through their highly novel "basic" technological approach. Such NTBFs are welcome challengers to the existing technological paradigms of large firm industry leaders. The innovative activities of such firms, rather perversely, may represent at once the best opportunities for rapid growth through the development of new high value technologies, while running the highest risk of total failure. While all high-technology industries will possess both "niche" and radically innovating firms, Figure 7.1 implies that differing high-technology sectors will possess such firms to differing extents. As argued above, the biotechnology industry, for example, is more likely to contain a higher proportion of radical innovators, with product developments further from the point of break even, while software firms are often "followers" in technological terms and may become self-sufficient almost from the point of formation. These different potentials imply a need for differing policy responses in terms of the types and amounts of assistance required by various new high-technology small firms.

Beyond the theoretical point made above that definitions that do not add clarity to our understanding are more damaging than no definition at all, definitions that do not allow for quite sharp functional differences between members of the defined class are dangerous in a functional sense. In such circumstances, a universal policy towards high-technology small firms might be appropriate to one sub-group within the general class, but damaging to other sub-categories. Indeed, in extreme circumstances, a generalised solution might not benefit *any* of the constituent sub-groups. The advantage of understanding detailed differences between various sub-groups of a broad generalised category is the facility it contains to allow policy makers to know when their actions are benefiting the whole group, or when additional measures are needed to assist the specific problems of a given problem sub-category. Here, a useful analogy with education might be adopted in which a population of generally high-technology small firms from different industrial sectors might be compared to the students in a given school year. While a general programme of teaching may be in place, special remedial teaching might be provided for small groups of students, or individual students, when and where appropriate. Industrial policy towards high-technology small firms might also accept this principle by, in addition to a general policy towards high-technology small firms, providing additional appropriate assistance in cases where *specific* sectoral needs, emanating from function differences, are detected.

This book has produced enough evidence to indicate that there *are* strong functional differences between new firms in the three high-technology industrial sectors of this study. While individual policy analysts can review the results and form their own opinions on how to react to such obvious differences, a number of observations are readily suggested by the results. First, it is generally clear that the biotechnology firms of this study have very strong and individual characteristics. The existence of most of these firms at the "basic" end of the model in Figure 7.1 has produced a number of extreme results, including, for example, extremely high levels of R&D spend and exotic input and output linkage relationships. External investors in such firms must understand that, unlike many other forms of high-technology production where lead times on product development can be protracted, this sector operates on the research horizon. Consequently, most biotechnology investments must be *extremely* long term in nature, involving great patience. Although there are good strategic reasons why a developed nation should possess a healthy population of new biotechnology firms in order to keep a national "foothold" in a clearly important world technology of the future, the need to support such developments financially may not be conversant with the needs and responsibilities of current financial markets (particularly in the United Kingdom where short-termism is rife). In such circumstances, there may be a strong case for public investment in cases (as discussed in Chapter 3) where a gap exists between public sector funding and private sector provision as a technology is developed from a basic scientific discovery towards the marketplace.

While private sector providers of investment capital may cite their shareholders' needs and the requirement for a good return on investment when refusing to fill such a gap, they *should not* then object if public money is advanced to meet the need that they were not prepared to accommodate. Medium- to long-term support by public institutions of technologically promising new enterprises is a realistic option, possibly on a non-profit-making basis, run on a true venture capitalist portfolio approach to investment. Such an approach is worthy of experimental trials, and has been advocated for several years by this author (e.g. Oakey 1984). In this context, the current United Kingdom government's competition schemes (e.g. SMART) can be seen as a lottery with finite financial limits that implicitly *do not* involve any form of long-term and *continuing* commitment. Funding continues to remain a barrier to high-technology small firm formation and growth, and is most acutely apparent in the funding of new biotechnology firms where considerable "front end" funding is (and will be) required in advance of any return from product sales (Figure 7.1; see also Oakey et al 1990).

The case of electronics firms in the survey provides an example of the need to understand the variable technologies that entrepreneurs can propose as a basis for a new firm formation. As Figure 7.1 implies, new electronics firms can be based on a range of technologies that are occasionally radical, but may also be of a "me too" nature. Various degrees of whole or partial "copying" or "following" using what are by then core technologies available publicly (e.g. semiconductors) allow often "niche" products that may or may not have mass market potential. Perhaps here in these rather heterogeneous conditions, a further fundamental shortcoming of policy, common to some extent in all areas of high technology, is seen at its starkest. Of equal importance to the willingness to invest in new industrial ventures is the need for expert knowledge on what are the technological proposals that have merit and when investment is extended, where should be its application, when invested in the firm (e.g. between R&D, production or marketing). Such applications will vary, depending on the degree to which the firm is developing radical or established technologies. Clearly, R&D support is essential for radically innovating firms, while assistance with marketing might be more appropriate for a firm competing in established "me too" markets.

To a large degree this problem of assessing the technological potential of a firm is a key determinant of the extent to which external investors are willing to advance investment support. Indeed, in the 1980s, United Kingdom banks oscillated from a position of unwillingness to lend in the early years, to excessive lending to many subsequently unsuccessful ventures by the end of that decade. Rather perversely, the errors of both these extreme examples had a single root cause, which was an inability to lend on the technological merits of the case which, had this ethos been followed, would have led to more firm funding in the early 1980s, and less by the end of the decade, with greater overall long-term effect in terms of survivals. The funding of firms should be judged mainly on the technical merits of the proposed activity which in turn influences its national and international market potential. The presence of such merit is a necessary condition to which other existing and/or subsequently acquired skills of the founders can add sufficiency. Thus the ability to judge such technical merit should be given to relevantly skilled engineers and not a bank employee or an accountant who has undergone a one-day technology appraisal course (Deakins and Philpot 1994).

The software firms of this study provide a good example of an instance where R&D is much concerned with "D" represented by development work to refine software products around a well established core technology. Here, more than other sectors, marketing and strong management approach is essential in conditions where product differentiation is not sharply defined and processes are very competitive. Many of the conditions that influence the operation of firms in this sub-sector of high-technology production are not especially high technology in nature and potential investors should appreciate that such firms are more likely to succeed in terms of price and service than because of the discovery of a new technologically innovative product that can be patented to yield intellectual property protected super profits to the firm. As noted above, software production in particular is not generally a vehicle for large and fast returns on investment.

The broader picture

Seen from a "free market" viewpoint, a logical reaction to the problems of new high-technology small firm innovation and growth, identified by this book, might be to argue that the three sectors of this study are not worth supporting if they cannot survive in competitive markets. It could be further argued that the high-technology industries to which they contribute should either survive on the basis of large firms, or not survive at all in the United Kingdom. However, apart from the actual and potential employment that respectively has been, and might be, provided by NTBFs, they will also make a contribution to a wider sophisticated high-technology industrial base through innovations within their own sector, and to other sectors with which they supply products and services. It is now well established that new high-technology small firms can provide substantial contributions to sectoral innovation (Rothwell 1982).

Freeman (1986) has also noted that new high-technology industries are potential catalyst sectors which not only grow internally but through their novel innovations radically improve the efficiency of other sectors. All three of the sectors of this study have such wider strategic potential. Examples abound of ways in which innovation in these sectors radically influences other sectors in cases where, for example, biotechnology products influence agriculture and medicine, electronic semiconductor-based products radically affect the design of automatic machinery, and software products improve the efficiency of all forms of product and process control in both the industrial and service sectors. Thus the high-technology new firms of this study, and the sectors to which they contribute, are key strategic industries that *will* play a large part in determining the industrial success of nations beyond the year 2000. These industries are of such critical importance that their survival and growth should not be consigned to the vagaries and anarchy of the "free market". Governments charged with the maintenance of national industrial bases should ensure that a conducive industrial environment exists for the survival and growth of such enterprises, either by persuasion of the private sector to meet the funding needs outlined above, or through the provision of such facilities when private sector market failure occurs. Hopefully, this book has shed detailed light on a number of specific problems that confront different new high-technology firms at formation and in the early years of life. However, in terms of the United Kingdom, although there has been a growing body of evidence on the problems of new technology-based firms since the early 1980s, there has been a dearth of effective policy measures designed to ameliorate these identified bottlenecks to growth. Without a strong government response aimed at addressing such problems, academic researchers into the problems of NTBFs will be reduced to "re-inventing the wheel" as, with the passing of time, we aridly rediscover problems that were not addressed when they were identified in the past. Such a state of ossification would not be conducive to either academic or industrial progress. Without prompt action, there is a strong possibility that this unwelcome state of affairs will continue for the foreseeable future.

Bibliography

Birch, D. L. (1979) The Job Generation Process, Working Paper, MIT Program on Neighbourhood and Regional Change, Cambridge, Mass.

Breheny, M. and McQuaid, W. (1987) High Technology UK: the development of the United Kingdom's major centre of high technology industry, in M. Breheny and W. McQuaid (eds.) *The Development of High Technology Industry*, Croom Helm, London, pp. 297-354.

Child, J. (1972) Organisational structure, environment and performance, in *Sociology*, Vol.6, pp. 1-22.

Cooper, A. C. (1970) The Palo Alto experience, *Industrial Research*, May, pp. 58-60.

Cyert, R. M. and March, J. G. (1963) *A Behavioural Theory of the Firm*, Prentice Hall, Englewood Cliffs, NJ.

Deakins, D. and Philpott, T. (1994) Comparative European practices in the finance of New technology entrepreneurs: UK, Germany and Holland, in R. P. Oakey, (ed.) *New Technology-Based Firms in the 1990s*, Paul Chapman, London.

Denison, E. F. (1967) *Why Growth Rates Differ*, Brookings Institute, Washington DC.

Farness, D. H. (1968) Identification of footloose industries, *Annals of Regional Science*, Vol.2, pp. 303-11.

Freeman, C. (1982) *The Economics of Industrial Innovation*, Frances Pinter, London.

Freeman, C. (1986) The role of technological change in national economic development, in A. Amin, and J. Goddard, (eds.) *Technological Change and Industrial Restructuring*, Allen & Unwin, London.

Glasmeier, A. (1985) Innovative manufacturing industries: spatial incidence in the United States, in M. Castells (ed.) *High Technology, Space and Society*, Sage, Beverly Hills, CA and London, pp. 55-80.

Greenhut, M. L. (1956) *Plant Location in Theory and Practice: The Economics of Space*, University of North Carolina Press, Chapel Hill.

Hall, P. (1962) *The Industries of London*, Hutchinson, London.

Harvey, D. (1973) *Explanation in Geography*, Edward Arnold, London.

Keeble, D. (1994) Regional influences and policy in new technology-based firm creation and growth, in R. P. Oakey, (ed.) *New Technology-Based Firms in the 1990s*, Paul Chapman, London.

Keeble, D. and Kelly, T. J. C. (1988) Regional distribution of NTBFs in Britain, in *New Technology-Based Firms in Britain and Germany*, Anglo-German Foundation, London.

Kelly, T. J. C. (1986) Location and spatial distribution of high technology in Great Britain: computer electronics, unpublished PhD thesis, Department of Geography, University of Cambridge.

Kuhn, T. (1970) Logic of discovery or psychology of research, in I. Lakatos. and A. Musgrave (eds.) *Criticism and the Growth of Knowledge*, Cambridge University Press.

Luttrell, W. F. (1962) *Factory Location and Industrial Movement*, NIESR, London.

Mansfield, E. (1968) *The Economics of Technical Change*, Longmans, London.

Markusen, A., Hall, P. and Glasmeier, A. (1986) *High Tech America*, Allen & Unwin, London.

Marshall, A. (1890) *Principles of Economics*, Macmillan, London.
Martin, J. E. (1966) *Greater London: An Industrial Geography*, Bell, London.
Mason, C. (1983) Some definitional difficulties in new firm research, *Area*, Vol.15, no.1, pp. 53-60.
Mason, C. and Harrison, R. (1994) The role of informal and formal sources of venture capital in financing of technology-based SMEs in the United Kingdom, in R. P. Oakey, (ed.) *New Technology-Based Firms in the 1990s*, Paul Chapman, London.
Morse, R. S. (1976) *The Role of New Technical Enterprises in the US Economy*, Report of the Commerce Technical Advisory Board to the Secretary of Commerce, January.
Murray, G. and Lott, J. (1992) Have UK venture firms a bias against investment in technology related companies ?, *Babson Entrepreneurship Conference*, July, INSEAD, Fontainebleau, France.
Oakey, R. P. (1981) *High Technology Industry and Industrial Location*, Gower, Farnborough, Hants.
Oakey, R. P. (1984) *High Technology Small Firms*, Frances Pinter, London.
Oakey, R. P. (1985) British university Science Parks and high technology small firms: a comment on the potential for sustained industrial growth, *International Small Business Journal*, Vol.4, no.1, pp. 58-67.
Oakey, R. P. (1991) High technology small firms: their potential for rapid industrial growth, *International Journal of Small Business*, Vol.9, no.4, pp. 30-42.
Oakey, R. P. (1993) Predatory networking: the role of small firms in the development of the British biotechnology industry, *International Small Business Journal*, Vol.11, no.4, pp. 9-22.
Oakey, R. P. (ed.) (1994) *New Technology-Based Firms in the 1990s*, Paul Chapman, London.
Oakey, R. P. and Cooper, S. Y. (1989) High technology industry, agglomeration and the potential for peripherally sited small firms, *Regional Studies*, Vol.23, no.4, pp. 347-60.
Oakey, R. P., Rothwell, R., and Cooper, S. Y. (1988) *The Management of Innovation in High Technology Small Firms*, Frances Pinter, London.
Oakey, R. P., Faulkner, W., Cooper, S. Y., and Walsh, V. (1990) *New Firms in the Biotechnology Industry*, Frances Pinter, London.
Popper, K. (1965) *The Logic of Scientific Discovery*, Harper Torch Books, New York.
Premus, R. (1982) *Location of High Technology Firms and Regional Economic Development*, Staff study prepared for the use of the sub-committee on monetary and fiscal policy of the Joint Economic Committee, Congress of the United States (97th Congress, 2nd session) US Government Printing Office, Washington DC, 1 June.
Roberts, E. B. (1991) *Entrepreneurs in High Technology*, Oxford University Press.
Rothwell, R. (1982) The role of technology in industrial change: implications for regional policy, *Regional Studies*, Vol.16 no.5, pp. 361-9.
Rothwell, R. and Zegveld, W. (1981) *Industrial Innovation and Public Policy*, Frances Pinter, London.
Rothwell, R. and Zegveld, W. (1982) *Innovation and Small and Medium Sized Firms*, Frances Pinter, London.
Rothwell, R. and Zegveld, W. (1985) *Reindustrialisation and Technology*, Longman, Harlow.
Semple, E. C. (1911) *Influences of Geographic Environment*, Holt and Company, New York.
Solow, R. M. (1957) Technical change and the aggregate production function, *Review of Economic Statistics*, Vol.39, pp. 312-20.
Speigelman, R. G. (1964) A method of analysing the location characteristics of footloose industries: a case study of the precision instruments industry, *Land Economics*, Vol.40, pp. 79-86.
Thomas, M. D. (1975) Growth pole theory, technological change and regional economic growth, *papers of the Regional Sciences Association*, Vol.34, pp. 3-25.

Thompson, C. (1988) Defining high technology industry: a consensus approach, *Prometheus*.
Thwaites, A. T. (1978) Technological change, mobile plants and regional development, *Regional Studies*, Vol.12, pp. 445-61.
Townroe, P. M. (1971) *Industrial Location Decisions*, Occasional Paper No.15, Centre for Urban and Regional Studies, University of Birmingham.
Wilson, H. I. M. (1993) An inter-regional analysis of venture capital and technology funding in the United Kingdom, *Technovation*, Vol.13, no.7, pp. 425-38.
Wise, M. J. (1949) On the evolution of the gun and jewellery quarters in Birmingham, *Transactions IBG*, Vol.15, pp. 57-72.
Wood, P. (1969) Industrial location and linkage, *Area*, Vol.2, pp. 32-9.
Woodward, J. (1965) *Industrial Organisation: Theory and Practice*, Oxford University Press.

Index

Acorn Computers 2
acquired rights 36
acquired firms
   innovation in 113
   numbers of R & D staff 70
   value of acquisition to 108–9
acquirer
   location of 108
   previous links with 108
acquisition
   and growth 105–7, 110–11
   criteria 107–9
   for funding 6, 103–5, 117
   prevention of 100–1
   target for 105–6
   see also independent survey firms; survey firms
advertising 95
agglomeration economies
   difference in location 83–4
   traditional industries 80, 89–90
Apple Computers 3
autonomy 103–4, 112, 117

barriers to entry 14, 26, 116, 120
break even point 119, 120, 121
bureaucracy 104, 112
business plan 52–5

Cambridge Science Park 3, 13
conferences 96
customer
   diversity 89–90, 91–2, 100
   international base 92–3
   patterns 89–90
   single firm 91–3, 100, 116–17
   single industry 90–92, 100, 116–17

defence industry 2, 3
determinism
   and technology 11–12
   in science 9
   principles of 9
   relevance of 11
   theory development 9–10
   see also entrepreneurs; firm's formation
employment
   funding growth 59
   in R & D 68–70
   problems 77–80
entrepreneurs 13, 14
   see also technology
environment
   competitive 99
   of firm 11
   see also founder
equity
   sale of 104–5
   shares 50–2
European Science Parks 2–3
exhibitions 96
exports 93, 98, 99

firm's formation
   definition of 29–30
   funding of 5
   motivation for 103
   problems 30
   sectoral difficulties 14
   technological determinism 12–14
   see also funding; models
founder
   former environment of 67, 119
   previous employer 18–19, 26, 115
   links with 21–22
   see also geographical influences
foundation stimuli 15–16
free markets 1–2
Fujitsu 2
funding
   at formation 46–7, 115–16
   for specific projects 58–60

future growth 60–1
gap 62
near market exploitation 34
R & D requirement 71, 80
subsequent to formation 47–8, 50–2, 57, 116
see also acquisitions; employment; government; investment

geographical influences 19–21
government
  funding 59–60
  interest of 4
  schemes 124
  The Enterprise Initiative 59–60
  SMART awards 59–60, 124
  subsidy 2
growth
  background to 1–3
  evidence of 1, 6
  link with sales and marketing 95–6, 97–8
  strategies
    aggressive 44
    and business plan 52
    risk-averse 43, 44
  in 1980s 2
  see also acquisition; employment; funding; models; production

high-technology industry
  agglomerations 89
  definition of 117–18
  misconceptions 3–4

ICL 2
incubator organisation 21, 103
independence
  financial 43, 51, 53, 54, 62
  of firm 67–8, 111
independent survey firms
  acquisition attempts towards 110–11
  relevance of acquisition to 109–12
in-house training 78–80
INMOS 2
innovation results 125–6
intellecutal value added 41, 83–4, 85, 87
investment
  acquisitions 103, 104, 109
  of formation and growth 43, 44, 45, 46–7, 54–5, 61

personal 43, 48
R & D 65, 116
by UK banks 125
see also venture capital
investors
  external 5
  in biotechnology 17
  management structure 51–2
  see also equity shares; investment

joint ventures 77

labour shortages 77
lead times
  and funding 45, 61, 62, 66, 119, 120
  near market exploitation 34
  variable 14, 119
location
  of customer 92
  of firm 5–6
  of suppliers 82–4
low-technology firms 5, 65–8, 68, 82

management
  structure 16–17, 51–2
  techniques 116
  technology choice 11, 12
mark ups 40
marketing
  agreements 98–9
  expenditure 95
  see also sales and marketing
markets
  international sales 90, 100
  see also niche market
Massachusetts Institute of Technology 13
materials
  imported 86
  input 84–7, 87
  input and output relationships 116–17
  procurement 81–4
  technology 6
models for new formation and growth 30–3
Morgan Cars 12

near market exploitation problems
  early spin off 34
  lack of support 33–4
  see also funding gap; lead times
niche market 94, 117, 121
niche production 100–01, 122

pay back period 14, 65, 80
personal freedom 15–16
post-formation
  external funds 50–2, 57
  product mix changes 41–2
  profits 48–50, 62
post-foundation product 35–6
premises
  quality of 22–3
  sectoral differences 23–5
  special needs 26

product
  additional 41, 42
  behaviour 29–35
  innovation 66–7
  life cycle 41, 42
  new ideas 15
  technology origins 35–6
  see also unit profit margins
production
  future growth areas 42–3
  see also niche production
profit
  margins 82–3, 88–9
  potential 67
  projected 60
  unit 45
  year end 55–6, 57–8, 62
  see also unit profit margins
projects see funding
promotion methods 94–5, 100
purchase linkages 82–4, 99, 100
purchasing patterns 84–9

research and development
  costs 71–2, 73, 116, 120, 124–5
  in-house 68, 73, 75, 80
  inputs 3–4
  licences 77
  linkages 72–7
  see also employment; funding;
    in-house training; sub-contracting;
    unit profit margins
Route 128 13, 82

sales
  income 49
  from initial technology 37–8
  linkages 82–4, 99
  profit contribution 55–6, 62
  unsolicited 93, 94

  visits 95, 97
sales and marketing
  resources 95–8
sectoral evolution 118–19
semiconductor industry 4, 113, 124
Silicon Valley
  firm formations 13, 20
  growth of 2, 3, 82
  markets of 89
  materials 94
  sub-contractors 87
  suppliers 84
Sinclair Research 2
single founder firms 16
skills shortages 78
staff
  sales 94
  working full time 95–7
Stanford University 2, 13
strategy
  founding 30–3
  industrial 9–14
  optimal formation 66–7
  ownership 104
  see also growth strategies
sub-contracting
  production 87–9
  R & D
    for external agencies 72–3;
    to external agencies 74
  see also suppliers
suppliers
  import sources 86–7
  interaction with purchasing 84
  international 100
  linkages 84–6
  local 81–2, 84, 85, 88
  sub-contractors 88
survey
  design 6
  implications for policy 122–5
  implications for theory 117–22
  process 6–8
  sectoral groups 4
survey firms
  acquisition in 105–7
  acquisitions achieved by 109–10
  initiating technology variations 36
  origin of founder 17–21
  R & D commitment 68–72, 80
  structure of 16–17
  see also independent survey firms

technical information
  links, 72, 75, 76
  sources of 74–6
technical service to customer 93
technological change 11
technological potential 124–5
technology
  leading edge 41
  market launch of founding 36–8
  product 35–6
  proven technology 116, 119, 121
  selection of 11–12
  see also unit profit margins
Thomson CSF 2
traditional industry
  abandonment of 2
  decline of 1, 11
  state subsidy of 2
training *see* research and development

unit profit margins
  of founding product technology 38–41
  link with R & D 68
US development 2–3

value added 83, 85, 87, 88, 92, 93
  see also intellectual value added
venture capital 51, 52, 60, 62, 103, 122

year end *see* profits

159883

This book is to be returned on or before
the last date stamped below.

18 FEB 1997

-7 MAR 1997

03 JUN 1997

CANCELLED
19 NOV 1998

LIBREX

OAKEY

159883

LIVERPOOL HOPE
THE BECK LIBRARY
HOPE PARK, LIVERPOOL, L16 9JD